THE ILLUSTRATED ENCYCLOPEDIA OF
SCIENCE AND THE FUTURE

Volume 17

Editorial Staff

Consultant Editor
John Ruck
Editors
Mike Bisacre
Peter Way
Martin Preston
Deputy Editor
Eden Phillips
Art Editors
Andrew Sutterby
Sue Walsham
Features Editors
John Birdsall
Jonathan Elphick
Ron Hyams
Noel Keywood
Lloyd Lindo
Clive Gregory
Gregor Ferguson
Alison Oldham
Editorial Secretary
Helena Ahmed

Designers
Peter Burt
Steve Chilcott
Dave Copsey
Paul Morgan
Phil Wildman
Marie Hemming
Simon Head
Clive Hayball
Richard Gliddon
Colin Truss
Technical Visualiser
Chris Mynheer
Senior Picture Researchers
Julia Calloway
Kate Parish
Picture Researchers
Jackum Brown
Louise Callan
Tessa Politzer
Marian Pullen
Marylinn Zipes

Miranda Innes
Mira Connolly
Mirco Decet
Flavia Howard
Picture Secretary
Angela Willenburg
Indexer and Proof Reader
Judith Devons
Picture Clerk
Amice Pilch
Production Executive
Robert Paulley
Production Controller
Patrick Holloway
Production Secretary
Linda Mifsud
Cover Design
Bob Burrows
Introduction
Martin Preston

Contributors

Ian Ridpath *Science writer*
Geoffrey Howard B.Sc. (Eng) (London) *Freelance technical journalist*
Nigel Henbest M.Sc. (Cambridge) F.R.A.S. *Astronomy consultant to the New Scientist*
John Williamson B.Sc. (Bradford) *Editor of Communications Engineering International*
Andrew Garrard Ph.D (Cambridge)
M.E. Horsley Ph.D (Sheffield) *Group Manager Fuel Technology, Portsmouth Polytechnic*
Anthony W. Robards Ph.D.,D.Sc. (London) *Reader in Biology, University of York*
Denis R. W. May Ph.D (Surrey), B.Sc. hons (London) *Technical Director, J. E. Hangar & Co. Ltd.*
Professor J. L. Cloudsley-Thompson M.A., Ph.D. (Cambridge), D.Sc. (London) Hon. D.Sc. (Khartoum) *Professor of Zoology, Birkbeck College, University of London*

Dr. Christine Sutton *Science Editor, New Scientist*
Eleanor Honnywill *(retired secretary)*
David Baker Ph.D., Dip. Ast. (J.S.C.), F.B.I.S. (Texas) *Technical Information Consultant Space Operations and Strategic Weapons Systems*
B.A.J. Ponder Ph.D., M.A., M.B., B.Chir (London) M.R.C.P. (Cambridge) *Cancer Research Campaign Fellow and Honorary Consultant Physician, Institute of Cancer Research and Royal Marsden Hospital*
W. P. Jaspert *Writer*
Donald Longmore F.R.C.S. (Edinburgh) M.B.B.S. (London) *Consultant Clinical Physiologist, the National Heart Hospital, London*
Robin Keeley B.Sc. (London) *Principal Scientific Officer, Metropolitan Police Forensic Science Laboratory, London*
Mavis Klein B.A. (Melbourne) *Psychotherapist in private practice, clinical member of the International Transactional Analysis Association*
J. D. Garnish B.Sc., Ph.D. C. Chem, M.R.S.C. *Manager of the U.K. Department of Energy Geothermal R and D Programme*
Dr. Ronald Zeegan F.R.C.P. *Consultant physician and gastroenterologist, St. Stephens and Westminster hospitals, London*
Peter Curran M.A., F.R.C.S. *Opthalmic Senior Registrar, Westminster Hospital, London*
Dr. A. D. Evans F.C.P. *Consultant Virologist, Public Health Department*
Bernard Ireland *Author of over seven books. Journalist and Illustrator, Specialist in Ship Research and development*

Cover Picture
Mike Peters

British Library Cataloguing in Publication Data
Way, Peter
  Insight.
  1. Forecasting—Dictionaries
  I. Title
  303.4'9'0321   CB158

ISBN 0 86307 069 8 (set)
ISBN 0 86307 086 8   Vol. 17

Printed in Great Britain by Ambassador Press.
Bound in Great Britain by Cambridge University Press

This edition published 1983

©Marshall Cavendish Ltd. MCMLXXX, MCMLXXXI, MCMLXXXII
58 Old Compton Street, London, W1V 5PA

All rights reserved. No part of this publication may be reproduced, stored in any retrieval system, or transmitted in any form or by any means, electronic, mechanical, photocopying, recording or otherwise, without the prior permission in writing of the Publishers.

THE ILLUSTRATED ENCYCLOPEDIA OF
SCIENCE AND THE FUTURE

MARSHALL CAVENDISH

NEW YORK · LONDON · TORONTO

CONTENTS Volume 17

Frontiers

Flesh-eating plants	2201
The X-ray eyes	2246
Earth's magnetic personality	2274
Man's origins : the biochemist's story	2290
Venus : inferno in disguise	2313

Electronics in Action

Hello Mr. Chips!	2206
Wired for sound	2302

Medical Science

The life supporters	2280
The big headache	2308
Cold cures : not to be sneezed at	2318

Machine Technology

Technology in the pits	2212
Road building : the way ahead	2240
France's high-speed train	2268
Forging ahead	2285

Technology Tomorrow

Cell of the century?	2324

Military Technology

The bear in the air	2218
Battle of the airwaves	2229
The people killer	2296

Resources

Soya : food of the future?	2262
Leather goes from strength to strength	2330

Sports Technology

The featherweight fliers	2224
Science serves an ace	2234

Forensic Science

Shot in the dark?	2252
Exposing the forger's art	2257

Frontiers: Botany

Flesh-eating plants

Though innocent looking and often strikingly beautiful, 400 of the world's flowering plant species have an ugly secret—they eat meat. Bugs, ants, flies, mice and even small birds, lured by the sweet scent of sugary secretions, tumble into these plants' lethal traps. Once caught, the prey may be crushed by leaves shaped into fleshy lobes, enveloped by sticky tentacles, drowned, or simply starved to death before being digested by the plants' acidic juices.

Trapping devices enable highly specialized plants to thrive in extreme terrains, such as acid bogs, mountain tops, or other areas that are particularly poor in nutrients. But while the meat-eating habits of some these plants are quite bizarre, the phenomenon is more than just a freak of nature. Botanists are gradually discovering that various degrees of meat eating may be remarkably common throughout the whole plant kingdom.

The leaves of the Venus flytrap, *Dionaea muscipula*—best known of the carnivorous plants—are shaped into prisons for insects. Each trap consists of two lobes, like a pair of cupped hands. Insects brushing against hairs on the surface of the plant's leaves trigger them to slam shut. The insect is then digested over a day or so. Then the trap reopens—unless the meal is so big that the whole leaf simply collapses with exhaustion!

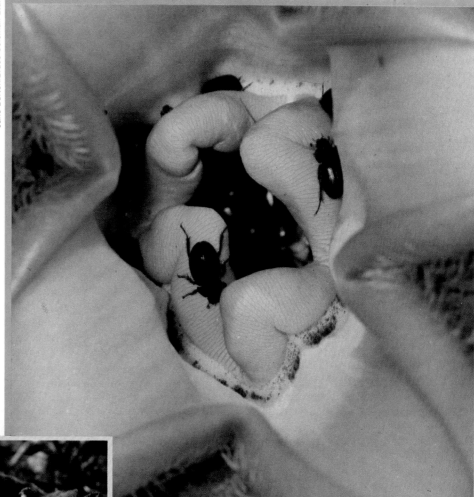

Above The beetle trap flower *Hydnora* (family *Hynoraceae*). A close-up of the interior of the flower shows a rich haul of trapped beetles scuttling over its fleshy inner lobes.
Left Beetle-eating plant *Hydnora,* with small dung beetles. Prey are attracted to the plant by the pungent rotting odour that it is able to produce from special glands. This remarkable, deadly plant is native to East Africa.

A lesser known cousin of the Venus flytrap is the Waterwheel Plant, *Aldrovanda,* which grows in freshwater pools. It has dozens of small underwater traps with which it catches small insect larvae and crustaceans.

The bladderwort, *Utricularia,* also has underwater or underground traps. Hairs on the bladder wall pump out water to keep the inside of the trap under partial vacuum, so that when the trap-door opens the unwary victim is swept inside on a mini surge of water. As quickly as it opened, the tiny door slams shut, imprisoning the struggling animal, which eventually dies of exhaustion.

Genlisia is also in the bladderwort family —the *Lentibulariaceae*—but its trap resembles an animal intestine. Its branched underground leaves are shaped into hollow tubes,

Frontiers: Botany

each ending in a small sealed bowl or *utricle*. Slits in the tops of the tubes enable insects to pass inside the trap, but once inside they are flushed away on a stream of water. Escape is barred by bunches of downward-pointing bristles, and as the victims pass deeper into the trap they encounter deadly arrays of digestive glands. Eventually the remains accumulate in the utricle, which but for the lack of an anus, bears an astonishing resemblance to an animal's digestive system!

Another carnivore of the plant world is the sundew, *Drosera,* which has leaf blades covered in a dense mat of sticky, tentacle-like hairs. Each tentacle secretes a gooey mucilage—which also contains digestive enzymes. An insect brushing past a tentacle near the edge of the trap is caught by strings of mucilage. In its struggle to free itself the animal brushes against other tentacles. Each tentacle is touch-sensitive, and bends over the animal to daub it with yet more mucilage. The more the prey struggles the more it becomes enmeshed, until the whole leaf curls up and envelops the animal.

Drosophyllum and *Byblis* employ similar —but immobile—tentacles to catch insects. In fact the *Drosophyllum* trap is so good that in Portugal the leaves are hung up in houses to act as a kind of natural flypaper.

In all these sticky traps the digestive glands are stimulated by chemicals released from the animal. The mucilage also has detergent-like properties which help the digestive juices wash into the small breathing holes along the insect's body—to speed the prey's death and eventual digestion.

Another trapping device is the water-filled trap of plants such as *Nepenthes* and *Sarracenia*. Their special leaves develop into elaborate bowls—or pitchers—which fill up with rainwater to drown prey. Insects are caught in abundance, but mice, lizards, and even small birds have come to grief in the giant pint-sized pitchers of *Nepenthes.*

The prey is lured to the rim of the pitcher by an irresistible combination of sugary droplets, bright red colours, and enticing scent. Once perched on the rim, insects lose their footing on the ribbed surface and tumble down into the pitcher. In *Sarracenia* the sweet droplets also contain an intoxicating liquor which makes the insects 'punch drunk'

Above Dragonfly caught in the leaf of a Venus flytrap. The only natural site where these plants grow is a sandy strip of bogland in North Carolina, USA. The lack of nutrients in the poor soil is compensated by the plant's fleshy diet, which also includes beetles and mites. *Left* Fly trapped on a sundew *(Drosera rotundiflora).* Note how the glandular tentacle hairs bend over the prey, gluing it with mucilage.

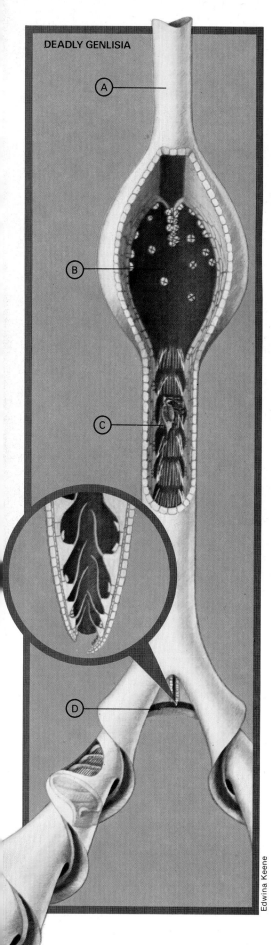

and more likely to stagger into the pitcher. Escape from the pitcher is thwarted by waxy cells, shaped like the downward-pointing tiles of a roof, which line the pitcher walls. Bit by bit the insect sinks down into the deadly bath of rainwater and digestive enzymes at the bottom of the pitcher.

Deadly Cobra

In the Cobra Plant, *Darlingtonia californica*, flying insects which have become imprisoned inside its leafy bowl look upwards and, seeing daylight, try to escape—only to collide with the plant's leafy hood and fall down with exhaustion.

The digestive juices of carnivorous plants are very similar to those of carnivorous animals—in fact the milk-curdling enzymes of the butterwort are used by Laplanders instead of rennet from calf intestines for making cheese. These plant juices also tend to be very acidic, and contain a number of enzymes for breaking down meat. The combined action of acid and enzymes rapidly dissolves the soft parts of an insect, producing a rich fluid that is absorbed by the plant.

Scientific techniques can be used to detect the digestive enzymes in a carnivorous plant's trap. One of the simplest methods is to lay a strip of developed colour photographic film over the plant trap. Protein-digesting enzymes then eat through the gelatine layer of the film, marking it with large holes. Other enzymes can be identified using colouring or fluorescing reagents, which can then be pinpointed under a microscope with the aid of a fluorescence detector.

The secretion of enzymes into plant traps builds up gradually over long periods. But with the Venus flytrap and butterworts the enzymes are stored in microscopic bags inside the digestive glands. When specific chemicals from the prey stimulate the gland, the bags burst open and the enzymes are squirted onto the trap surface along with an acidic solution.

Many plants capture animals, but not all are truly carnivorous. Quasi-carnivorous plants may rely on micro-organisms to break down their prey if they lack the ability to produce their own digestive juices. Studies at the Royal Botanic Gardens in Kew, UK, show that a supposedly conventional carnivorous plant—the Sun Pitcher, *Heliamphora*—may in fact be an imposter, since it has no digestive glands or enzymes. In all other respects, though, *Heliamphora* is a model carnivorous plant.

Many plants have leafy pitchers or water-filled leaf bases, and many collect organic debris—including dead insects. For example, the pitchers of *Dischidea rafflesiana* clearly make use of their organic dustbin by grow-

Left The strange form of the leaf trap of *Genlisia*. It has a cylindrical foot stalk (a) and a hollow bulb (b) from which issues a cylindrical neck (c). At the end it has a slit-like mouth (d). Tiny water animals are sucked towards the plant's trap in the current set up by water-pumping hairs. The inside of the plant is designed with rows of hairs in the mouth and up the neck pointing inwards to act as cone-shaped valves. The tiny 'lobster pot' lets animals enter but does not let them escape.
Right Sundew *Drosera*.

Left The pitcher plant, *Sarracenia*. Insects are attracted both by the sweet secretions from the rim and its red colours—which are the same shade as rotting meat. Once caught, escape is prevented by waxy angled scales. *Below* Huge stalked *Triphyophyllum* glands.

ing a root down into the rotting material.

The 'ant plants' also have quasi-carnivorous pitchers. These curious plants hire a protection squad of ants to fend off the attacks of other herbivorous insects. In return, the ants receive board and lodging in the form of hollow shelters and nectar-like food. But recent experiments have shown that the plant obtains another benefit—in *Hydnophytum* the ants actually feed their host plant with dead insects.

In one ingenious study small fruitflies were fed on a diet containing molecules tagged with a radioactive label. The radioactive flies were left near the plant to await collection by the ants. The ants took the flies back to the plants and lined the walls of the pitchers with them. Using a sophisticated Geiger counter, the insect remains were then traced. They were found to pass through the walls of the pitcher into the plant. The ants thus provide the means of entrapment for the quasi-carnivorous *Hydnophytum* as part of a remarkable symbiotic relationship between plant and animal.

Not all pitcher plants are flowering plants. The leafy liverworts are a group of delicate moss-like plants which grow on the sides of trees or on stones in humid areas. Living out of contact with the ground, they are particularly vulnerable to shortages of water and minerals, and have evolved microscopic pitchers to capture and hold rainwater. The shape of these pitchers varies widely, from the helmet-shaped *Frullania* to the more exotic slipper forms of *Colura* and *Pleurozia*.

Ingenious trap-door

When microscopic organisms pass into the pitchers of *Colura* and *Pleurozia* they brush past an ingeniously simple but fiendishly effective trap-door at the entrance of the pitcher, which closes behind its tiny victims like a letter-box flap. Whether the pitcher actually captures its prey is not known. However, it has been noted that when the pitchers are dry they collapse like a crumpled paper bag, and on re-wetting they expand so rapidly that any nearby organism is soon sucked in.

In the past 50 years only one new genus has been officially recognized as being carnivorous—the rare and very beautiful *Triphyophyllum* from West Africa. *Triphyophyllum* is a liana—a climbing plant common in tropical rain forests. At the height of the rainy season some of its leaves develop spectacularly large tentacle glands—the largest known glands in the plant kingdom. The tentacles secrete large droplets of sticky fluid, which catch flying insects and spiders.

Triphyophyllum grows in extremely mineral-poor soil and the development of the insect-catching glands shortly before the rainy season enables the plant to obtain the nutrients it requires to shoot up and carry its flowers and seeds high into the treetops.

Another plant genus worthy of recognition as being carnivorous is the crucifer family—including mustard and cress. These

2204

produce seeds which develop a thick sticky coat of mucilage during their germination. Recent research has confirmed that insect larvae, nematodes, and many micro-organisms have a fatal attraction to a substance which diffuses out from crucifer seeds such as those of the common weed, Shepherd's Purse *(Capsella bursa-pastoris),* and traps insects by their mouths.

The rich supply of enzymes in the seed then break down their captives into easily absorbable food, which passes through the seed coat into the embryo seedling. Such fleshy food supplements certainly may not be essential to the seed's diet, but the extra minerals provide a welcome boost to the seedling.

The hairs of plants such as the Red Catchfly *(Lychnis viscaria)* and some saxifrages are also highly sticky insect traps. And sticky hairs covering the leaves of the garden *Petunia* even contain a lethal poison which kills off insect visitors. These secretions often increase during flowering, and the hairs probably evolved to protect the foliage and flowers from insect pests.

Similarly, the sticky hairs of the common sundews, *Drosophyllum,* may also have evolved for protection, before the plant turned to meat eating. Supporting this idea is the fact that the structure of the hairs of digestive glands of these plants all conform to the same basic three-cell design, or *trichome*.

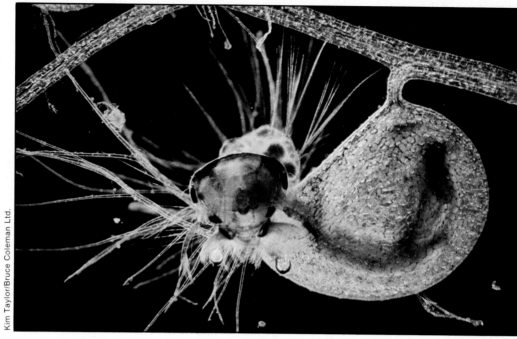

It is becoming increasingly clear that carnivorous plants are not the rarities they were once thought to be. For more and more plants are being found to be at least slightly carnivorous—even certain tomatoes *(Lycopersicon hirsutum)* and potatoes show carnivorous tendencies! Careful observation has shown that small insects—many of them common pests such as the spider mite and greenfly —become entangled in the sticky hairs of the shoots of these plants. The dead insect remains then rot and it is quite likely that they pass through the hairs into the plant. As digestive enzyme has been discovered in the potato hairs, flesh eating is entirely plausible.

More exciting still, the development of hairs is controlled by a single pair of genes, so that the characteristic of hairiness could be bred into hairless plants already in cultivation. Indeed, a breeding programme is already underway at the Plant Breeding Institute, Cambridge, UK, to cross the hairy tomato with a more familiar variety.

If carnivorousness can be bred into crop species it would provide the plant with a natural insecticide combined with a useful supplement of nutrients—without the expense and pollution risks that accompany the use of artificial chemicals. The day of the Triffids may be nearer than we think.

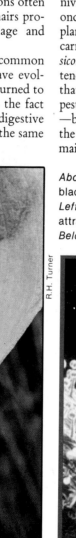

Above Lesser Bladderwort *(Utricularia minor)* bladder containing an engulfed mosquito larva.
Left Lords and Ladies *(Arum masculatum)* attracts prey with a stalk smelling like meat.
Below Sticky hairs trap and digest an aphid.

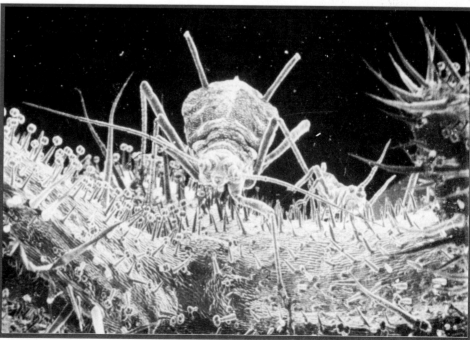

Electronics in Action: Education

Hello Mr Chips!

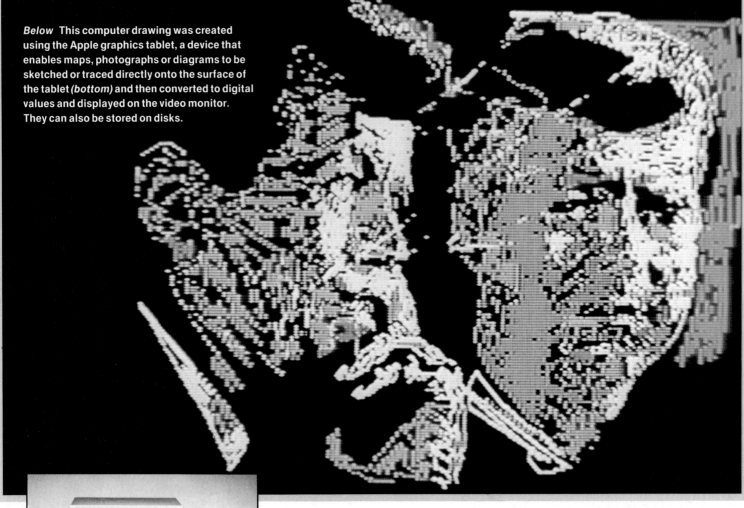

Below This computer drawing was created using the Apple graphics tablet, a device that enables maps, photographs or diagrams to be sketched or traced directly onto the surface of the tablet *(bottom)* and then converted to digital values and displayed on the video monitor. They can also be stored on disks.

The classroom has already absorbed many of the educational aids of the technological age—from language laboratories and overhead projectors to calculators and video recorders. Now a microcomputer on every school desk is a possibility for many developed countries before the end of the century. Costing as little as a standard typewriter, a microcomputer in every home is also a distinct possibility—thereby calling into question the necessity for traditional schooling methods.

Early development of the computer was centred entirely on the central main-frame computer, with schools linked directly via terminals. Plato, the United States computer-based education system, is of this type. Lessons and course material are stored in the central computer and are accessible through television terminals operated by students via a keyboard and a touch-sensitive screen.

In Europe and the USA, the recently developed microcomputer has increased the range of applications enormously. About the size of a portable television, these personal computers can function as either a terminal or independently. They incorporate their own keyboards for input and screen for output, with the central processing unit or micro-chip providing the memory. Colour, sound and graphics can all be incorporated.

In Viewdata systems like the British Prestel, a television set and a simple adaption to the telephone line provide access to all kinds of computerized information, much of it useful to schools and colleges. Pages of data are displayed on the television screen, with Prestel acting as an intermediary between the information providers who create the pages and the customers who are automatically debited a small amount for being able to gain access to them.

Surprisingly, children of below average ability were found to excell using the Prestel information service for searches. Since the information is paid for by the user per page ordered, the children had to learn to order only relevant data. Less able children were able to get to the root of their search by the shortest means, while teachers and brighter children tended to order far more information than they needed.

'Network Nation'

'Network Nation' is a term coined in the USA to describe an integrated system which would include microcomputers and other attachments in addition to a Viewdata system. A simple device called a MODEM (MOdulator/DEModulator) already allows direct contact between microcomputers and the telephone system. The telephone handset rests on the MODEM and digital information is translated into, or decoded from, sound pulses. A standard MODEM can translate 30 characters per second, but some versions are capable of translating up to 1,200 characters per second.

Prestel are currently collaborating with the British Council for Educational Technology on a project known as *telesoftware*, which will use the Prestel system for the direct transmission of programs from one computer to another. The receiving computer can be ready for use immediately, or the incoming software programs can be stored on cassette or disk attachments.

Computer literacy—that is, learning how to operate computers—is one educational use of the new technology. A strictly logical process, operating a computer first requires fluency in the language for which the device was designed. Computer languages are not difficult to use. They amount to an adaptation of normal word-and-figure communication designed to be understood by an electronic system that is basically thousands of on-off switches.

Experiments have shown that young primary school children are capable of quickly learning enough to construct simple programs in BASIC (Beginners All-purpose Symbolic Instruction Code), one of the most common computer languages.

Computers are now available in a form that is attractive, or 'user friendly' to students of all ages. Software has been designed with immediate appeal, making use of pictorial representations called *interactive graphics*—often on a colour screen—which form quickly in response to the operator. These graphics may be charts, cartoon characters, or pictures of all kinds.

Computer-Assisted Learning—following courses controlled by computer without having to understand programming—is an alternative strand of development. A good example of the simplest form of Computer-Assisted Learning is the range of learning devices produced by Texas Instruments. These include spelling, arithmetical and logic games in the form of a small pre-programmed module which is much like a pocket calculator.

The most sophisticated of these learning aids is 'Speak and Spell', which stores up to 240 words. It has four available levels, and new clip-in memory modules are being produced to extend its range into such difficult

As personal computers become more inexpensive and versatile, computer-based classroom learning for all students is spreading rapidly. The Apple *(below)* makes learning fun.

Electronics in Action: Education

areas as irregular spellings and homonyms.

Students work through programs at their own pace, with the computer adapting the route through the work according to how well certain tasks are performed. A student begins by selecting a program from those offered on the display screen. The range depends on the contents of the memory.

The computer's synthetic voice asks questions and the student types the replies on the integral keyboard. The voice either congratulates a correct answer or suggests another try—perhaps at a lower level—in response to a mistake. Often in game format, computers are appealing to children, and can be a valuable aid for acquiring literacy and numeracy—particularly for slow-learning or foreign-speaking children.

Because the computer is patient and does not express annoyance, children with learning difficulties are often happier working with a machine and less afraid to reveal their mistakes to it. Remedial teachers have discovered that children's concentration and levels of achievement have improved con-

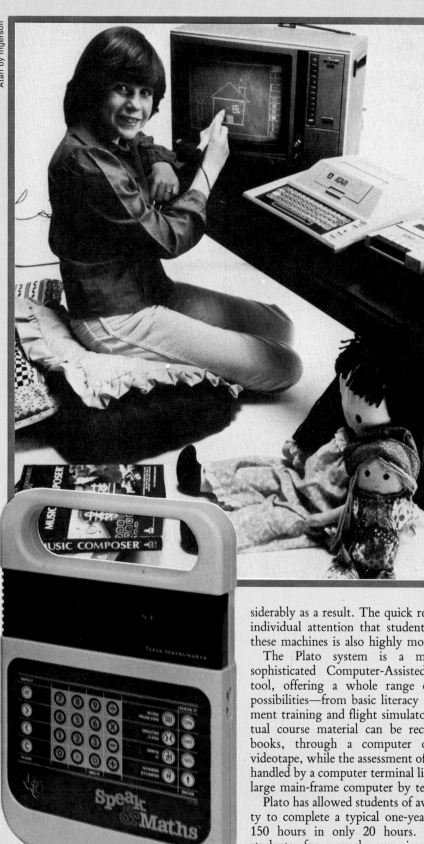

Right The Atari 400 personal computer, about the size and cost of an electric typewriter, plugs into a TV. Here, graphics are displayed as they are drawn on the screen with a light pen.
Below Texas Instruments learning aids.

siderably as a result. The quick response and individual attention that students get from these machines is also highly motivating.

The Plato system is a much more sophisticated Computer-Assisted Learning tool, offering a whole range of learning possibilities—from basic literacy to management training and flight simulators. The actual course material can be received from books, through a computer or from a videotape, while the assessment of progress is handled by a computer terminal linked to the large main-frame computer by telephone.

Plato has allowed students of average ability to complete a typical one-year course of 150 hours in only 20 hours. Navigation students, for example, can simulate operation of a ship's controls and steer a course that changes according to their responses —all in the classroom using a computer pro-

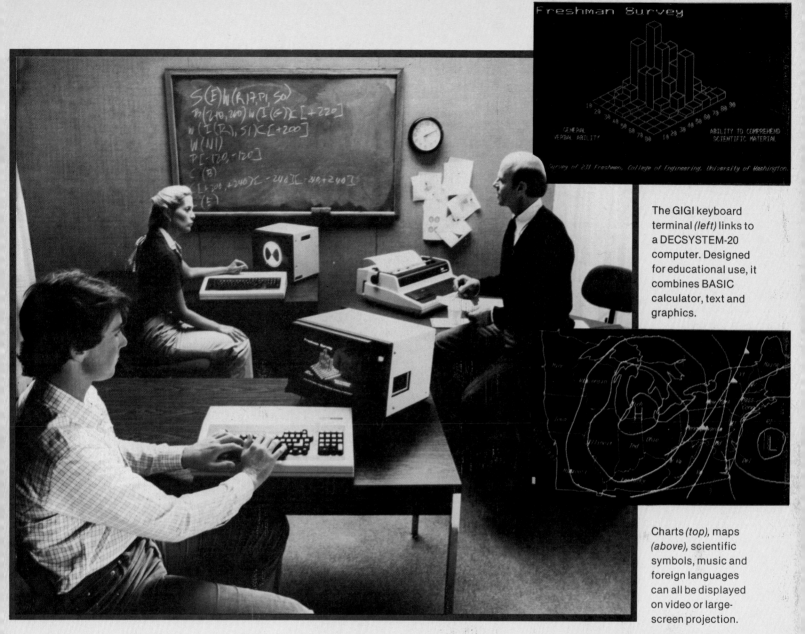

The GIGI keyboard terminal (left) links to a DECSYSTEM-20 computer. Designed for educational use, it combines BASIC calculator, text and graphics.

Charts (top), maps (above), scientific symbols, music and foreign languages can all be displayed on video or large-screen projection.

gram of a radar display. The student works at his or her own speed, getting an immediate and personal response from a course that can be constantly updated and redesigned on the main computer.

Apart from systems like Plato, there is a burgeoning of software—programs—and computer add-on hardware—printers and graphics tablets—to expand the capabilities of the microcomputer.

Educational programs

Programs are available to teach basic literacy and numeracy, accounting, design, languages, aspects of medicine, maths, sciences, technical drawing—in fact any part of the educational curriculum. For those who lack manual dexterity computers can execute a painting. Even music can be included with a synthesizer. Students think up a 'solution' and the computer expresses it.

One computer tool of particular importance in education is the graphics tablet. This is a small flat board equipped with a special stylus enabling drawings to be produced on the VDU screen. Dotted points can be joined with perfect lines, and images can be moved and reduced on the screen. Drawings, designs, maps, histograms and plans can all be produced and manipulated with the minimum of effort. A typical example of the graphics tablet's use is to enable students to design printed circuit boards. Complex plans can be altered with just a stroke of the pen, and the design can be linked to a program which actually tests whether it will work.

In Britain, a scheme to install at least one microcomputer in every secondary school was recently unveiled—accompanied by a complementary scheme to train teachers in the use of computers, plan curriculum changes and provide computer advice to schools through regional centres.

In France, too, there has been an ambitious programme to buy 10,000 microcomputers—about four in every school. France's centrally directed educational system has eased the setting up of a project using a computer language based on the French vocabulary, and a system whereby schools can exchange material. Among the main difficulties in expanding its project is teacher training which, along with software provision, lags behind the installation of microcomputers.

In the USA, educational programs are mainly for use in the home or school. The material tends to be didactic and directive, unlike the more interactive material in use in the UK, for example.

Electronics in Action: Education

Philips LaserVision

Above The Philips LaserVision disc system. A laser beam shines onto the underside of the disc and is reflected back through the lens *(left)* to produce sound and pictures on a TV.

SOFTWARE GLOSSARY

ARRAY A series of items (data or information) arranged to form a meaningful pattern.

BASIC Widely used microcomputer language. An acronym of Beginners All-purpose Symbolic Instruction Code.

BITS BInary DigiTS. The 0s and 1s that make up the binary code computers understand.

BYTE A term to measure a number of bits, usually eight bits to a byte.

CHIP A tiny piece of silicon which holds all the components that make up a microprocessor.

DECODE To interpret and determine meaning.

DISK A circular piece of magnetically coated material, onto which information can be stored.

EPROM Erasable Programmable Read-Only Memory.

FLOPPY DISK Flexible disk onto which computer data can be stored.

FOR . . . NEXT In BASIC, an instruction used for repetition of a sequence of program statements.

FORTRAN FORmula TRANslator.

GOSUB A BASIC command instructing the computer to go to a subroutine in a computer program.

GOTO A BASIC command which tells the computer to jump to another line in a program.

GRAPHICS CHARACTERS The name given to pictorial representation of data such as plotted graphs, engineering drawings and computer games.

HARDWARE A general term given to all pieces of electronic and mechanical devices which make up a computer system.

KILOBYTE A measurement of memory capacity: 1,024 bytes.

LANGUAGES Sets of words, symbols and corrections used to instruct the computer.

L.E.D. (Light-Emitting Diode) provides a simple display and consists of an electron tube which lights up when electricity is passed through it.

LOGICAL NETWORK A series of interconnected points linked by communications facilities.

LOOP A BASIC function referring to the repeated execution of instructions for a fixed number of times.

MEMORY That part of the computer where information is stored.

MODEM A device which connects a microcomputer to the telephone.

PEEK A BASIC command to read the contents of a byte.

POKE An instruction used in most versions of BASIC allowing integers to be stored in a specific place in a memory.

PRINT A BASIC command which tells the computer to perform a calculation.

PROCESSING Handling and manipulating computer data.

PROM Programmable Read Only Memory.

RAM (Random Access Memory) A memory chip in which data can be stored.

ROM (Read Only Memory) A memory chip which can only be read from and not written into.

ROUTINE A set of coded computer instructions used for a particular function in a program.

SCAN To examine stored information for a specific purpose.

SOFTWARE The programs fed into a computer, which make them perform the desired function.

SUBROUTINE A computer program routine that is translated separately, generally used in several computer programs or several times in one program.

VARIABLE A symbol whose numeric value can be changed at all times. It is used when writing programs.

Advances in educational technology are of most benefit to the disabled. There already exist control systems for the severely physically disabled, such as those produced by Possum Controls. Joy sticks, pressure pads, foot switches, light pens, and suck/blow tubes enable those with very limited movement to activate a variety of devices, including environmental control systems, typewriters and communicators.

In a typical communication system, a screen displays a choice of words or symbols with a quickly moving pulse of light passing each one. When the operator works a control, the pulse stops and the selected word is typed or displayed on a screen. Such a system is slow, usually limiting the user to one activity at a time.

With the microcomputer, however, the possibilities become much greater. Mechanical switches are replaced by a microcomputer connected to a microprocessor. The signals that the computer receives can be made to control other machines, display text on a screen for viewing or editing, or make a drawing. Using computer programs, the disabled pupil who is unable to speak or write can respond to multiple-choice questions and is less dependent on the teacher.

Storing information

The software is also available for educational establishments to organize testing, student records, time-tabling and other administrative matters on microcomputers. But one area with which the small computer cannot yet cope is in storing the huge amounts of information required by libraries.

Video discs, due for launch in 1982 in the UK by Philips, are an ideal basis for an information-storage system. They can hold encoded material of any kind, and whole encyclopaedias can be stored on a single 30 cm (12 in.) disc and then displayed page-by-page using a freeze-frame technique.

Video discs incorporate a laser which beams a finely focused beam at a disc covered with a reflective surface containing millions of tiny pits. A light sensor deflects the disturbances that are created by the pits, and they are converted into an electric signal and fed into a television set.

Large libraries can obviously be served by expensive 'main-frame' computers. The British National Reprographic Centre for documentation is currently working on designs to serve the smaller or growing library. This will involve a number of microcomputer or word-processor work stations linked to a very large memory system. Information can be stored for computers on cassette tape, floppy disks (which are read by a special add-on disk reader) and hard disks, which are capable of storing vast amounts of information. Libraries of any size will certainly need a hard disk system. Eventually it may be possible to tie-in microcomputers to a global library, making all world knowledge instantly accessible at any time or place.

Word processors are a relatively new development, used mostly in commerce at present. But as well as adapting their large memory capacity for libraries, word processor systems are also likely to be used increasingly to design and update educational courses. Computers and processors can both be linked to daisy-wheel printers which are capable of reproducing words and images.

The microcomputer, more than any other educational aid, has wide-ranging implications for the future, not least for the traditional relationship between teacher and student. The idea of the teacher as dispenser of information is fast becoming obsolete. As new knowlege is rapidly discovered, data banks which can be revised constantly are better able to provide information for tomorrow's students. Teachers may instead become advisers, suggesting computerized information sources.

Many educators believe it vital that today's school children become as familiar with computers and other data bases as with books. Computer literacy and familiarity with the mechanics of information processing, storage and retrieval will be skills required of all students to exploit their abilities in the computer age.

Right The Atari 800 personal computer has a range of facilities for educational use. Its ability to display 16 colours is particularly useful for graphics. The Atari 810 disc drive in the foreground offers fast-access storage.

Machine Technology: Motor racing

Technology in the pits

For the 20th time, in the 1981 Italian Grand Prix at Monza, John Watson eased his Marlboro McLaren MP4 car through the 150 mph right hand bend known as the second Lesmo. Almost lazily the left hand wheels drifted towards the chequered kerbing as the Irishman realized he had entered a shade too fast. Suddenly, the red and white car snapped into a spin, speared backwards into the opposite guardrail, and was literally torn in half.

Without doubt, the accident was one of the worst seen in motor racing for several seasons, and that Watson could step unharmed from his wrecked car is testimony to the strength of current Grand Prix machines. But the accident also provided a vivid illustration of the shortcomings of the new breed of upper echelon racing cars. A year earlier Watson could simply have ridden out a slide over the kerbing and continued his race; but a change of regulations during the winter changed all that. Now the suspension movement of the cars is virtually non-existent.

Perhaps suprisingly, the reason for the ultra-hard suspension has little to do with mechanics. Instead, it is employed to optimize the cars' aerodynamics. In order to understand why this is so, it is necessary to step back into motor racing history. Nothing is new in motor racing, but it takes a shrewd designer to take an old concept and turn it into something that works well today as well.

As early as the 1920s a rocket-powered Opel appeared with stubby aerofoils protruding from its flanks. Then, as now, the idea was to generate downward pressure, or *downforce*, to keep the car's wheels on the road. It was an idea that failed to catch on. In the late 1960s, however, a young Californian engineer named Jim Hall reintroduced the aerofoil—basically an inverted aeroplane wing section that produced downforce instead of lift—on his successful Chaparral sports racing cars. Within a year of the cars appearing in Europe in 1967, virtually every Grand Prix team followed suit as they were gripped by 'wing fever'.

It must have been gratifying for Hall, although he never made any money from the profusion of aerodynamic aids. However, by 1969 his mind was on other things and the following year he produced a racing sports car—the Chaparral 2J—so advanced and so fast that it demoralized its opposition whenever it appeared. In simple terms, he had come up with a means of literally sucking the car to the ground.

Beneath the 2J's boxy exterior was a system of Lexan plastic *skirts* which acted as seals between car and track. To the rear, and in addition to the 7 litre Chevrolet engine, was a snowmobile motor connected to two fans. As the motor drove the fans, air under the car was extracted and the chassis was pulled closer to the track by the suction.

The 2J was banned almost immediately, but its legacy was to be felt in the motor racing world many years later. Over the following years racing car aerodynamics became of secondary importance, and the Grand Prix car became an object of steady development. Few could envisage another revolution such as the advent of the rear-engined-car. But during the 1977 season two cars appeared that were to change the face of GP racing. One made use of special aerodynamics, the other of a new type of racing engine.

Below Nothing is really new in motor racing. The 1981 Williams of former world champion Alan Jones echoes the 1928 Opel rocket car with its inverted aerofoils and rock-hard suspension *(below right)*.

Lotus is probably the most successful Grand Prix team of all time and with rumours of an all-new car for 1977, it became clear that the team was on the verge of a major comeback after two disappointing seasons. After months of research and painstaking development, the Lotus 78, designed by the brilliant Colin Chapman, was released upon a startled GP world.

In essence, the entire monocoque chassis of the 78 was to serve an aerodynamic function so that it would not simply be a bluff hull with a parasitic effect on the remainder of the components. To achieve this, the pannier fuel tanks and water radiators were mounted on either side of the slim chassis and featured upwardly curving undersides so that when they and the body panels were in position the sides of the car resembled an inverted aerofoil wing section.

Beneath the car a system of skirts separated the airflow around the car from that beneath it. In effect, the new car used Hall's skirt idea to control undercar airflow, but used wing section sidepods to generate downforce rather than a separate motor.

Above left Jim Hall's 1970 Chaparral 2J used fans and skirts to 'suck' the car onto the track. Colin Chapman's Lotus 79 *(left)* had skirts, but used an aerofoil-section underside as well to achieve a similar effect.

The second major development of the season was the advent of the turbocharged, or forced induction, engine in Grand Prix racing. Forced induction engines had fallen from favour in the 1950s, although the regulations governing successive GP formulae continued to allow them to participate. Up to 1977 few exploited this opportunity, but when Renault, the French motor manufacturer, decided to promote its image by entering Grand Prix racing, it took the bold step of designing a turbocharged V6 power unit.

The current rules equate a 1.5 litre forced induction engine with the more popular 3 litre normally aspirated units and when the Renault first appeared few believed it had any chance of success. But after some four years of intensive development, turbo engines have become the most feared on the circuits and there have been moves to have the equivalency changed to 1.4 or 1.3 litres.

Renault's V6 Gordini engine shuns the use of exotic materials. Its camshafts are driven by toothed belts and the early units used Kugelfischer fuel injection in conjunction with a Garrett AiResearch turbocharger. This employs a turbine which is driven by the engine's exhaust gases. In turn, it drives a compressor which forces air into the engine instead of simply letting the unit inhale of its own accord. With more fuel and air to burn the result is a considerable increase in power output. The chief disadvantage is the time lag between the driver opening the throttle and the turbocharger taking effect.

Turbo power

During its early life Renault's adventurous concept proved highly unreliable, but in 1979, when a twin turbocharged engine with greatly reduced throttle lag was mated to a ground effect chassis, Jean-Pierre Jabouille won the French Grand Prix in convincing style and both the French manufacturer and Ferrari (who now also field turbocharged cars) have since won regularly.

A turbocharged racing engine differs significantly from a turbocharged road car engine, but experience garnered from the intense competition of Grand Prix racing has taught Renault's engineers much about eliminating throttle lag and simplifying installations—and all this filters back to their road car programme.

In the year after the Lotus 78's stunning performances, several rival teams set about making up lost ground—but Chapman had already built a new car. This was the Lotus

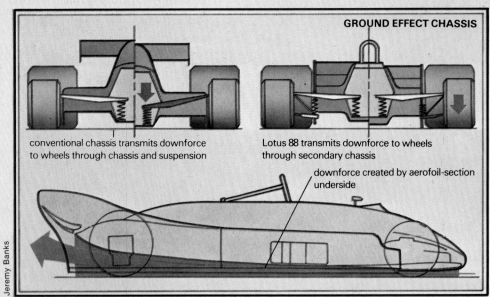

GROUND EFFECT CHASSIS

conventional chassis transmits downforce to wheels through chassis and suspension

Lotus 88 transmits downforce to wheels through secondary chassis

downforce created by aerofoil-section underside

Jeremy Banks

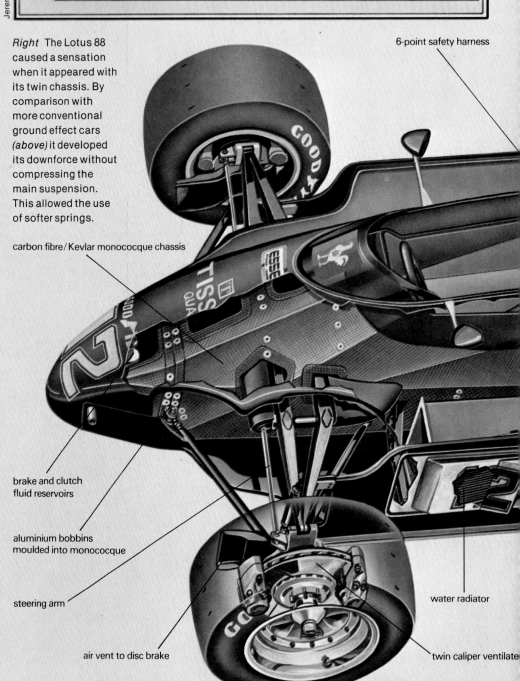

Right The Lotus 88 caused a sensation when it appeared with its twin chassis. By comparison with more conventional ground effect cars *(above)* it developed its downforce without compressing the main suspension. This allowed the use of softer springs.

- 6-point safety harness
- carbon fibre/Kevlar monocoque chassis
- brake and clutch fluid reservoirs
- aluminium bobbins moulded into monocoque
- steering arm
- air vent to disc brake
- water radiator
- twin caliper ventilate

Machine Technology: Motor racing

79, a machine of true aesthetic beauty yet, when devoid of its sleek bodywork, of almost frail appearance. Gone were the pannier fuel tanks of the 78 and in their place were full length, aerofoil-section sidepods.

The fuel was now all carried in a single tank behind the driver and the venturis created by the sidepods accelerated the undercar airflow and ensured that it exited cleanly. With special skirts once again sealing the air chamber beneath the car, by rubbing on the road surface, a low pressure area was created, sucking the chassis to the ground and providing impressive levels of grip.

Throughout 1978 the Lotus 79 was the car to beat, winning the Driver's championship for Mario Andretti and the Constructor's title for Chapman's team.

When he saw how effective the 79 was, Brabham designer Gordon Murray began to consider means of improving his BT46 design. Jim Hall had used an auxilliary engine to suck out undercar air; Murray reasoned that it should be possible to achieve a similar effect by using an engine-driven fan. When the Brabham was driven quickly, the fan would rotate fast, sucking out the air beneath the car and providing the drivers with even more grip than the Lotus.

Lauda won the Swedish Grand Prix with this car, but immediately after, the protests flooded in. The main controversy centred on whether or not the fan constituted a moveable aerodynamic aid, something forbidden by the regulations.

Ultimately the Brabham went the way of the Chaparral 2J and once again the Grand Prix world settled down to a period of entrenchment, designers either seeking ways of improving levels of grip or simply struggling to understand ground effect technology.

A major improvement came with the introduction of skirts which could slide up and down as the car rode over bumps, for by maintaining contact with the track surface at all times the skirts ensured maximum ground effect and led to even higher cornering forces being generated.

LOTUS 88B FORMULA 1 RACING CAR

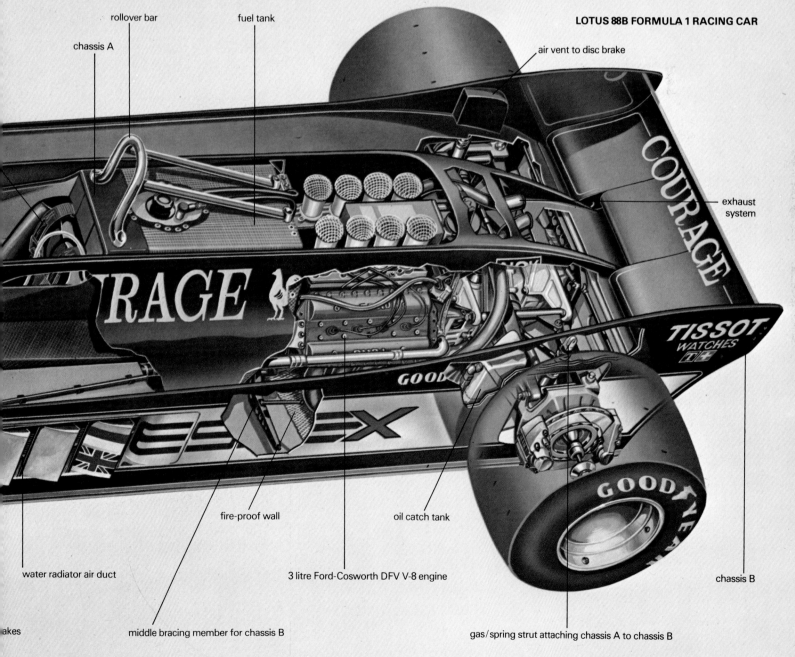

Machine Technology: Motor racing

Left 1981 and the skirts on the Marlboro MP4 no longer touch the track. A carbon fibre chassis makes it very crashworthy! Tyrrell's 6-wheel P34 *(bottom left)* was a modest success but doomed to ultimate failure, while the Williams 6-wheeler *(centre left)* trades lightness for handling and aerodynamics.

By the end of 1980 FISA (Fédération Internationale de Sport Automobile), the sport's governing body, had become alarmed by the speed at which current cars were cornering and decided that as from the first race of 1981 sliding skirts would be illegal. Moreover, new regulations demanded a ground clearance of at least 60 mm in order to reduce the level of ground effect created.

Even before the sliding skirt ban, Chapman had been looking ahead and had concluded that the ideal ground effect car would be one with two chassis, one carrying the driver and running gear, the other the bodywork. When the ban on sliding skirts was imposed, cars began to run with much stiffer suspension, to keep their undersides as near parallel to the ground as possible, so the idea seemed even better.

The main chassis could be run with fairly soft springs, giving the driver a more comfortable ride, and the second chassis could be flexibly attached to it. Thus at rest the car complied with the 60 mm ground clearance ruling, while at speed the aerodyamic downforce acting on the secondary chassis forced it down on its own set of springs so that its rigid skirts ran much less than 60 mm from the ground.

Like Brabham in 1978, however, Chapman now found his car the subject of protest for contravening the regulations, with the result that it was rejected out of hand by FISA before it had even raced.

At the beginning of the 1981 season most Grand Prix cars resembled their 1977 counterparts, running fixed skirts and large aerofoils front and rear in an attempt to provide adequate downforce. But by the Argentine Grand Prix in March, Gordon Murray had found a loophole in the regulations. Basically, this took the form of a hydropneumatic suspension system which automatically lowered the ride height of the car once it reached racing speeds so that once again the 60 mm rule was circumvented.

Since the 1930s racing car straight-line speeds have risen little; the improvement in lap times that occurs year after year has simply been a result of phenomenal advances in cornering abilities.

The World Championship-winning Williams team is now studying means of increasing the top speed of its car without sacrificing cornering ability and to this end has produced a six-wheeled version of its FW07 car. Six wheelers are nothing new, even in modern times; in 1976/7 Tyrrell ran the interesting but relatively unsuccessful P34, while March Engineering had a brief flirtation with a six-wheeled design. Tyrrell opted for four small front wheels in the interests of reducing aerodynamic drag. But the large rear tyres resulted in a still high frontal area and the special 300 mm (12 in.) diameter front tyres were never sufficiently developed. The March used four rear wheels the same size as the front ones but ran into trouble, with the twin rear gearboxes failing to work in harmony.

Positive benefits

By no stretch of the imagination can the racing developments outlined above qualify as benefits. But the aerodynamic research data gained by racing teams breaks new ground and will ultimately benefit future fuel-conservative road cars. And the pioneering use of new materials and construction techniques, as exemplified by Marlboro McLaren's carbon fibre MP4, should not be overlooked. Perhaps the greatest benefit would result from the adoption of a formula based on fuel efficiency. This would oblige engine manufacturers to build power units that were efficient throughout the speed range rather than capable simply of producing high power outputs, and the techniques learned could soon be passed to the road car industry, with consequent benefits to each and every motorist.

Below Racing can still improve the breed! Renault's turbocharged V6 engine was the first to appear on the track, followed by Ferrari's example *(below)*. Racing experience here bears directly on road car design.

Military Technology: Aircraft

The bear in the air

According to Soviet military planners, defence begins at home. Not necessarily in concrete bunkers for officials and administrators assigned to the task of getting things back to normal after a major nuclear war, but rather in the most complex series of air defence measures ever mobilized by one nation. The Russians are totally committed to the defence of Soviet territory and that philosophy has been extended into space where even ballistic missiles may soon be vulnerable to laser weapons and ray guns. Their military thinkers believe Russian towns and cities should be screened so that no strategic weapon could ever get through—even in the event of a nuclear war.

The Soviet Union is unique in having an air defence structure capable of deterring concentrated attack at any level. It has not always been so. As recently as the 1950s, Soviet air space was virtually undefended. There was no long range radar, few fighters to deter intruders, and little co-ordination and control for those defence systems that did exist—mostly guns and makeshift rocket projectiles. Time and again, Soviet military forces watched helplessly as huge American B-36 bombers roamed high above the Soviet steppes with thermo-nuclear bombs.

Stalin's successors

US bombers posed a serious threat to Soviet Russia and brought the world very close to war—averted only by the death of Josef Stalin, who was replaced by a group unwilling to press ahead with plans for an invasion of Western Europe. But the bombers remained a constant threat. By 1955, the United States Strategic Air Command had nearly 1,600 long range and intercontinental bombers, each capable of dropping a multi-megaton bomb on cities deep within Soviet territory.

This was at a time when the intercontinental ballistic missile (ICBM) was only a vague hope and the biggest operational missile was the US Thor, then in the design stage, with a maximum range of 2,500 km. Nevertheless, based in Europe, the missile posed an additional threat over and above that presented by a seemingly invulnerable armada of multi-engined aircraft cruising through the stratosphere.

The single most important incentive for the massive build-up in ballistic missile forces on both sides was the increasing vulnerability of manned bombers, which were required to fly within several kilometres of the target. From this point the bombs would be released and glide to earth or to a point of detonation high in the atmosphere. Nevertheless, determined to rid themselves of the threat of a massive US bomber fleet over their territory, Soviet planners threw unprecedented effort into protecting the country by the defence of Soviet air space.

The Soviet National Air Defence Command (PVO Strany) became an independent element of the Soviet military structure in 1954. It now has 6,000 Russian and Warsaw Pact personnel (little short of the total number of US Army personnel), some 2,600 front line interceptors, more than 10,000 surface-to-air-missiles (SAM) and launchers, more than 8,500 anti-aircraft guns, and

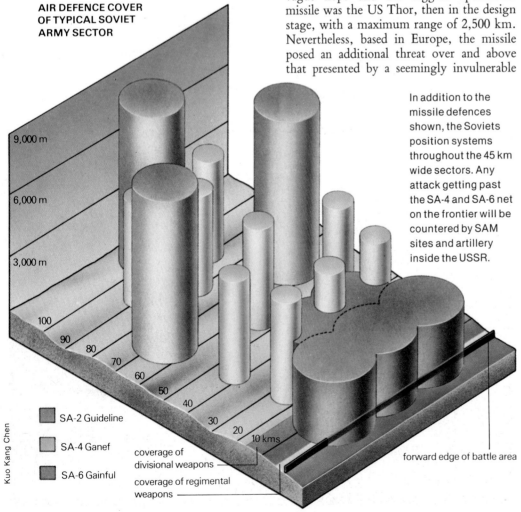

AIR DEFENCE COVER OF TYPICAL SOVIET ARMY SECTOR

In addition to the missile defences shown, the Soviets position systems throughout the 45 km wide sectors. Any attack getting past the SA-4 and SA-6 net on the frontier will be countered by SAM sites and artillery inside the USSR.

- SA-2 Guideline
- SA-4 Ganef
- SA-6 Gainful

coverage of divisional weapons
coverage of regimental weapons
forward edge of battle area

Kuo Kang Chen

flying bomber, while Britain's decision to rely on a nuclear deterrent based at sea rather than on a fleet of V-bombers, had wide repercussions, relegating the UK Vulcan force to low level penetration and nuclear strike in support of defensive measures should Russia attack the West.

By the end of the 1960s, Russia had effectively locked out the traditional force of manned bombers and demonstrated that it could quickly shoot down high flying aircraft intent on penetrating Soviet air space. This was accomplished by four separate branches of the PVO Strany, each set up to deal with a vital function of air defence.

Radar troops were made responsible for manning the several thousand ground, air and spaceborne radars essential for monitoring the frontier. Using the Tupolev Tu-126 Moss aircraft from 1973, these radar troops acquired instruments of radar detection that were soon to be copied in the West. Known as AWACS (Airborne Warning And Control System), they are essentially compact radar detection platforms 'orbiting' 12,000 m above the border and protected from attack by interceptors of the PVO Strany. The

Left SA-1 Guild surface-to-air missiles in the 1981 Red Square Parade in Moscow. Estimated to have been in operational service since 1954, SA-1 is now being replaced.
Below SA-3 missiles are used in countries outside the Soviet Union. This fixed version has been installed near Helsinki in Finland.

about 7,000 ground, air and spaceborne radar systems. In wartime, the Frontal Aviation Units would be integrated with the PVO Strany elements through the commander of ground forces.

U-2 spy plane

The enormous home defence problem set by Russia's long frontiers was met successfully only with the introduction of high performance missiles. The first, or SA-1 type, was responsible for shooting down an American U-2 spy-plane flying from Pakistan to Norway in 1960. This marked the turning point in strategic weapons. By 1967 the USA had reduced its manned bomber force to less than 600 aircraft and had a total of 1,700 ballistic missiles, mounted in underground silos or on board nuclear submarines. The enormous speed of these missiles was thought to render them invulnerable.

Increasing air defence reserves in the Soviet Union reduced the threat of the high-

Military Technology: Aircraft

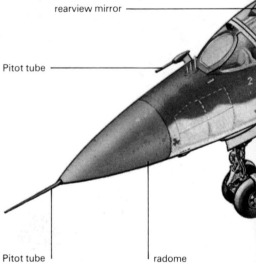

Left An opportunity for closer inspection of the MIG-23 Flogger occurred when six of these aircraft visited France and Finland during 1978.

MIKOYAN – GUREVICH MIG – 23S FLOGGER – B

- ferry tank
- dual frequence COM system and antenna
- air intake
- rearview mirror
- Pitot tube
- Pitot tube
- radome

AWACS aircraft are the Soviets' long range eyes to detect aircraft movements—and they have at least ten of them, easily identified by their large, circular antenna housings on top of the fuselage.

Advance radar warning

Ground-based Tall King radar units are used in Warsaw Pact territories for advance warning of enemy aircraft. Such warnings are passed to a central control and co-ordination facility where weapon systems are activated to shoot down the invader, either by manned interceptors or else from a fixed SAM site. If a SAM site is assigned to the interception role, a Fansong, Long Blow, Long Track or Pat Hand radar is slaved to the incoming target. Simultaneous missile tracking and control enable the operator to use a single source for both target acquisition and missile guidance.

From space, surveillance takes on a global challenge, with satellites equipped to pick up the tell-tale signs of a missile suddenly appearing from its silo or undersea location. Strident efforts are now being made to place greater responsibility on satellites for keeping track of aircraft movements over the country of origin. In this way, Soviet Air Defence Command would immediately be aware that a large formation had strayed from the host country's air space.

To shoot down uninvited aircraft, the PVO Strany operates anti-aircraft artillery troops. The most commonly used weapon, accounting for almost half the total 9,000 anti-aircraft guns, is the 57 mm S-60, with a maximum range of 12,000 m; it is a radar controlled towed gun and may also be used as an anti-tank weapon. Next in order of preference is the ZSU-23-4, a self-propelled gun system with quadruple barrels lethal to low flying aircraft. The vehicle has a range of 250 km and is fitted with tracks for cross-country use.

The Soviets have a tradition for never discarding a weapon system just because a new and improved version becomes available. The dated equipment remains in use in less critical areas, and this trend is seen to good effect in the anti-aircraft missiles used by the SAM troops. The first type deployed, the SA-1, is still in use, but since the late 1950s there has been a succession of different models, each designed for a specific task. The SA-2 Guideline formed the backbone of the SAM missile force for more than two decades, following its introduction in the early 1960s. It is effective against targets flying at 25,000 m.

The SA-3 Goa is a low altitude system mounted in pairs on a tracked vehicle, while extremely long range aerial threats are subject to the most powerful Soviet SAM, the SA-5 Gammon, which has a length of 16 m and a weight of 10 tonnes. This finned missile is brought to the vicinity of the

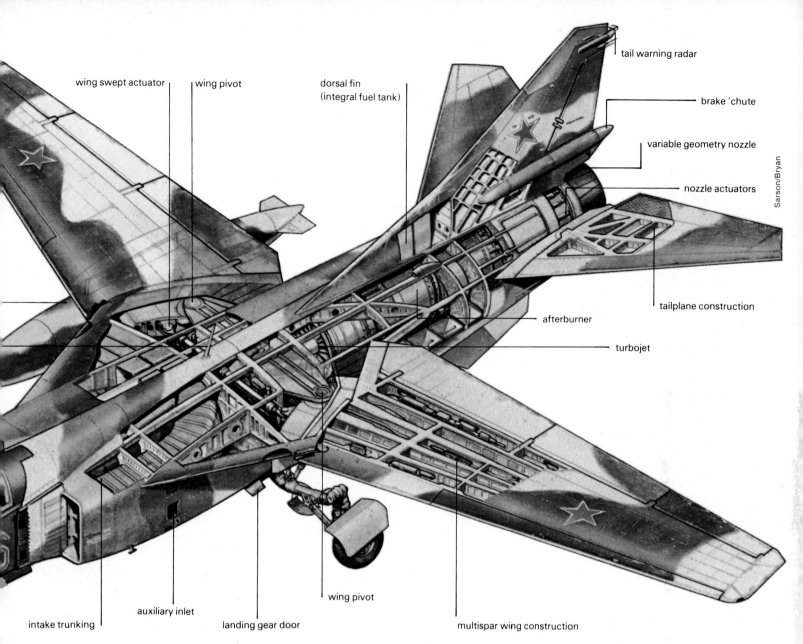

Above and *right* The MIG-23 Flogger swing-wing aircraft is by far the most important interceptor in the Soviet Air Force. Its armament includes one internal 23 mm twin-barrel cannon, in addition to various mixes of air-to-air missiles.

target, perhaps 250 km from the point of launch, guided by a Square Pair radar which then hands over control to the Gammon's own internal radar system.

Portable SAM launcher

With maximum ranges of 75 km and 55 km respectively, SA-4 Ganef and SA-6 Gainful missiles are employed on tracked launchers for battlefield air defence and protection over rear areas. Battlefield threats that get through the Ganef/Gainful net are countered by a shoulder-mounted SA-7

Grail, which can hit an aircraft at a range of 3 km. A derivative of the Grail has been mounted on tracked launchers and is effective only against the smallest aircraft at a maximum range of 10 km.

The SA-8 Gecko, SA-9 Gaskin and SA-10 are new missiles introduced during the late 1970s. Both SA-8 and SA-9 are self-propelled, the former with a range of 13 km, the latter with a shorter range but broader application. The SA-10 was, until the early 1980s at least, an enigma shrouded in secrecy and rumour. Best estimates are that the missile can attain an altitude of 9,000 m, has an active radar guidance unit and attains a speed of Mach 6.

A natural development of surface-to-air missiles occurred when ballistic rockets took over as the prime nuclear delivery systems. If the manned bombers could be stopped by SAM rockets, so too could the big ICBMs, or so was the theory in the early 1960s. But intercepting an incoming warhead, travelling at 25,000 km/h was made seemingly impossible when technical developments led to MIRV (Multiple Independently targeted Re-entry Vehicle) warheads capable of flying to separate targets after release from the missile's nose cone. Now, instead of finding one target the anti-ICBM missiles would have to contend with two, four, six, even ten separate incoming warheads per missile. Moreover, decoys released with each warhead confused ground radar into thinking each was a potential threat.

But Soviet concern outstripped the philosophy of deterrence, in which the vulnerability of cities and industrial sites supposedly preserves the peace since neither side would risk annihilation by retribution. Accordingly, PVO Strany developed an ABM (anti-ballistic missile) system with a rocket called Galosh. This has never publicly been exhibited, so that for the present it is seen as only four nozzles exposed by the open end of a container housing the missile.

Galosh is integrated with massive Hen

Right This version of the MIG-25 Foxbat is fitted with versatile camera installations and radar sensors, and is used for reconnaissance.
Below A Tupolev Tu-126, with distinctive AWACS (Airborne Warning and Control System), being shadowed by a US Phantom multi-role aircraft.

Above The ZSU-23-4 is a neat package of firepower with quadruple 23 mm barrels. It is self-propelled and is fitted with its own target acquisition and fire control radar.
Left The SA-9 is mounted on an amphibious scout car, and it is assumed that targets are acquired by radars in other vehicles.

House radars systems, which seek and identify incoming objects. Missile management is carried out with Dog House or Cat House radars which have a range of 2,750 km. These are located close to the Galosh silos, and together with Chekhov trackers provide the ABM system with an effective counter to a limited number of ballistic missiles.

The PVO Strany also provides the manned punch to destroy penetrating strike aircraft which, in the event of war, would seek to fly at a low altitude and at high speeds to targets deep inside Russian air space. To counter the

NATO nuclear strike element, more than 2,500 interceptors protect the borders and key sites along the entire length of Russia's territorial boundary. Moreover, an additional 1,000 interceptors from Warsaw Pact countries provide an initial buffer.

Interceptor protection

Most numerous in the inventory are the 800 Sukhoi Su-15 Flagon fighters, capable of Mach 2.3 and a combat radius of more than 650 km. Next are the 750 Mig-23 Floggers, just supersonic at sea level but with a maximum Mach 2.2 at altitude and an effective radius only a little less than the Flagon. Most famous of all are the 350 Mig-25 Foxbat fighters. Capable of speeds in excess of Mach 3, it can climb at 15,000 m per minute and fly 1,100 km to the war zone before releasing its missiles and returning to base.

The balance of nearly 700 aircraft in the air defence inventory is made up of a motley collection of comparatively old designs. By the early 1980s the PVO Strany had only 120 of their Tu-28P in service; the biggest fighter ever built, this aircraft has a maximum speed of Mach 1.75 and an incredible range of 2,900 km. It carries infra-red and radar guided missiles and is an all-weather fighter. Finally, the Yak-28P Firebar, a more recent design, is built in both interception and strike variants. It too carries missiles for the fighter role and has a range of 1,600 km.

By far the most important interceptor is the Mig-23 Flogger, which during the 1980s will replace Su-9, Su-11, Tu-28P and Yak-28P types. Further improvements to the Mig-25 will extend that aircraft's performance, which was considered to be disappointing when a Soviet air defence pilot defected to Japan and presented the West with a perfect model for examination.

One of the most significant areas of development is the Protivokozmicheskaya Oborona (PKO). This little known section is responsible for defence against space weapons and for collecting information on US spy satellites and electronic ferrets which swoop low from orbit and gather radio signals for analysis in the West. The PKO will probably assume greater importance when the development of space-based weapon systems reaches a peak around the end of the 1980s.

It has been known for some time that the Soviet Union has been working towards more sophisticated laser and particle beam weapons. To control and operate all the many separate tactical and strategic weapons effectively, satellites are an indispensable tool in guiding warheads accurately to their targets, in allowing military units to communicate, and in reporting accurate details of the enemy's movements.

To be the first to destroy in the early stages of an attack, satellites as vital as these would bring supreme advantage to the attacking side. Consequently, the Russians have developed a family of 'killer-satellites' capable of ascending rapidly from launch and detonating an explosive charge to disable enemy satellites. In 1980 a more refined concept emerged: the Soviets placed in space a battle station carrying several interceptors, each of which could be fired at its own target guided by infra-red sensors.

These types of space weapons must be considered as an integral part of the Soviet air defence system, in as much as they compromise the chances of an enemy pressing home an attack. In the final analysis, however, no screen is totally impenetrable and even behind this unique air defence operation Soviet leaders expect to suffer heavily should a major war break out.

SURFACE TO AIR MISSILE (SAM) SA-7 GRAIL

This simple infantry weapon is fired from the shoulder, and is lethal only against small aircraft. There is also a tracked version which has a more effective range. The missile itself runs the whole length of the launcher and is fitted with special flick-out fins.

Sports Technology: Model Aircraft
The featherweight fliers

What do airship hangars, half a football match, castor oil, a couple of cigarettes, balloons, Pirelli rubber, and Romanian salt mines have in common? The answer is indoor microfilm competition model aircraft. Man's preoccupation with the theory and practice of flight has led him along many paths, but few more unusual than the pursuit of duration records with model aircraft within the confines of a building.

Microfilm is the name given to the models used in the Fédération Aéronautique Internationale (FAI) class of indoor miniature aircraft—and it also applies to the aircraft's ultra-light, transparent covering. There are many classes of indoor model aircraft, ranging from tiny scale replicas to ornithopters (flapping wings), but microfilms are the ultimate class for world championships. In their devotion to keeping these featherweight aircraft airborne for a maximum period of time, competitors—often from rival countries—meet regularly to compete and to exchange advice and assistance.

The fascination for this highly specialized form of model flying owes much to the ethereal beauty of the aircraft, their thistledown lightness, and the amazing efficiency of the designs. Indeed the design, construction and flying of microfilm models is both an art and a science. The beauty of the models comes partly from the transparency and the frailty of construction, and partly from the design of the delightfully curved flying surfaces and propeller.

Microfilm models are not for the ham-fisted. They are among the most fragile artefacts, rivalled only by some of the most miraculous examples of natural design and engineering. These delicate craft perform best in a calm, almost motionless atmosphere, so competitions are usually held in airship hangars and Romanian salt mines—the only structures that offer the vast indoor areas required for long flights.

Above Preparing a microfilm aircraft for flight requires patience, skill and a little luck. Having wound the rubber motor to the correct number of turns —found largely by trial and error—the modeller attaches it to the fuselage and propeller, taking care not to damage the delicate craft.
Right The colours reflected by the film help to make the model visible and add to its beauty, but they also indicate whether the film has been correctly made.

Above A network of roof trusses, into which a model could easily become entangled, compounds the difficulties of microfilm flying. The danger is reduced by steering the models with balloons tethered to lengths of cord.

Today, the hobby has been developed to the stage where the FAI has had to limit flight durations and airframe lightness for models competing in the World F1D class. This action was necessary to keep competitions within reasonable time intervals and reduce the incidence of structural failure due to frailty. The maximum wing span for this class is 650 mm (25.6 in.) and minimum weight of the bare airframe is one gramme.

A flying model of this size limited to a frame weight of one gramme appears impossible, even to expert modellers. Yet such is the standard of competition that even one hundredth of a gramme could be critical when maximum duration is the aim.

Every part of the structure should be just sufficiently strong to withstand normal flying loads. Any excess of structural requirements will carry too great a weight penalty. Balsawood is selected for a minimum weight that will give just the right strength, so not only must the quality be uniform and high, but the way the wood is cut is crucially important.

Balsawood can be cut so that it is stiff and suitable for wing ribs, or flexible as in flying surface outlines. All wood is weighed before it is used, and the sizing of the structural components is critical. Usually the fuselage tube is fashioned from wrapped balsawood tubes, giving a rigid, lightweight structure capable of taking the compressive stresses of the wound rubber motor without distortion.

Depending on the wing rib construction, the flying surfaces and propeller are built on a jig; this not only ensures accurate construction, but also enables the modeller to make several copies of a model, each having the same dimensions. This is essential because competitors need to take more than one model to a contest, and preparation is much easier if the models are near identical.

Because flying conditions are so consistent, theoretical flying times can be predicted accurately. To achieve these parameters in practice, however, calls for the highest modelling and flying skills. Using Pirelli rubber to power the propeller, flying durations approaching 45 minutes—half a football match—are achieved. Despite these impressive flight times the total weight of the models is slightly more than two grammes —the equivalent of two or three cigarettes.

Microfilm modelling became practicable in the early 1930s after balsawood—an extremely light but strong building material —became widely available to modellers. Until then, spruce, bamboo, tissue, oiled silk and piano wire were the standard materials for building model aircraft. By combining balsa and microfilm techniques, flight times of more than 20 minutes were reached rapidly—but the 30-minute target remained elusive for many years.

Microfilm aircraft design did not change radically: much research was needed to perfect construction techniques, reduce weight and improve the power source. Eventually, the 30-minute barrier was exceeded, followed swiftly by the 40- and 45-minute durations for 'open class' models.

The microfilm covering

Making the microfilm covering requires great skill. The weight and texture of the film are critical for optimum flying performance. The technique is to prepare the film material as a liquid floating on water. Nitrocellulose dope is used as base material and to this is added castor oil or camphor, which acts as plasticizer and prevents the film from becoming too brittle. Then amyl acetate is used to thin the mixture.

In judging when the mixture is correct, experts are always guided by the reflected colours of the film when the liquid is poured onto the surface of water. Green-blue, purple and gold are the colours they aim for to give

an ideal thickness and stability of film that is not too sticky.

Domestic baths make suitable water trays for producing the microfilm, but dedicated enthusiasts make their own polythene-lined tank which is considerably larger than the film area required. A teaspoonful of the mixture, poured over the surface of the water, spreads towards the edge of the tank but never reaches it. Once the mixture and pour are correct, the film is attached to a simple, rectangular wooden frame—larger than the component on the model to be covered—by placing the dampened frame over the film. The film then adheres naturally to the wooden frame, on which it is lifted clear of the tank and left to drain and dry.

After a storage period to stabilize it, the framed film is lowered on to the component to be covered. The flying surfaces of indoor models are covered only on one side, but larger models often have their wings and tailplane covered on both sides. Care has to be taken in the covering operation, and with the handling of the components afterwards.

To prevent the wings from being floppy when fixed to the fuselage, bracing wires are attached to the wings, fuselage and to a king-pin above the wings. This additional rigidity also prevents the wings from warping and changing the flight trim.

To keep the model light, the bracing must be extremely thin. Either tungsten with a diameter of 0.031 mm or 0.020 mm in nichrome is adequate.

Propellers are built on jigs and must be balanced precisely to prevent the low-frequency vibrations that would cause the model to oscillate. One innovation that adds to the difficulty of building and flying microfilm models is the variable-pitch propeller—which weighs less than the filter on a cigarette. Already in use, these propellers compensate for the variations of power through the unwinding of the rubber motor. Some modellers use a computer to predict the performance of the propeller and motor throughout the flight.

Transporting a model to the flying venue can be a problem, particularly when travel-

Right Accurate scale drawings are essential for competition models because weights and dimensions are limited. But they also enable modellers to make identical craft so some can be kept as spares, in case one is damaged.

ling abroad. Packing the component parts of the model in a box is particularly difficult: due to its insubstantial structure, it cannot be fixed firmly to any packing. It has been known for a competitor to arrive at a World Championship meeting with six models, only to find them all seriously damaged. And microfilms cannot be repaired or rebuilt overnight, or brought up to flight perfection in a couple of test flights.

Buildings with cavernous, uninterrupted interiors are not plentiful, and microfilm models often have to be flown with restricted roof heights. This is possible only if the models are derated to limit climb. For indoor flying, the nearest to ideal conditions are probably in the Romanian salt mines at Slanic, where there is more than adequate ceiling height, minimal temperature gradient

Sports Technology: Model Aircraft

Above The ideal flight programme, in which a model climbs to optimum height and descends as the motor runs down, is achieved by measuring the precise number of turns on a mechanical counter. Another invaluable aid is a magnifier *(left)* for viewing fine detail, as well as the barely visible microfilm.

throughout that height and extremely calm air. In the USA and Britain, old airship hangars are popular for indoor meetings; these memorials to a bygone age of aviators are so large that they may even have their own 'weather' inside. A sudden cooling of the external temperature can cause precipitation inside and some strong downdraughts which may reduce flying time by half.

At Cardington in Bedfordshire, UK, the airship hangars built originally for the ill-fated British R 101 have been the venues for many of the Indoor Model World Championships. Affectionately known as 'the sheds', they are nearly 250 m long by 50 m high. Their cathedral-like interior is awe-inspiring, and many team members from visiting countries have more than doubled their previous best flying times simply because of the vast space available.

Despite the enormous volume of air, however, any violent movement by people in the shed can cause air waves sufficient to disturb a model in flight. Internal flying conditions are at their best when the external weather state is calm, warm and dry. Cold winds or squally showers cause less predictable flying and models might drift across the hangar, making towards the steel members of the structure. Assessing the correct trim of the model and the number of turns to apply to the rubber motor are also made more difficult if the temperature gradient fluctuates considerably; the aim is to climb the model to as near to the roof as is practicable without contacting any of the roof trusses.

Luck plays a part in some of the best times achieved, but tactics are nonetheless crucial. Being able to judge when the ideal conditions are about to occur and deciding which member of the team should fly are essential decisions for the team manager. Only the two best times—from a possible six flights—are aggregated to determine the

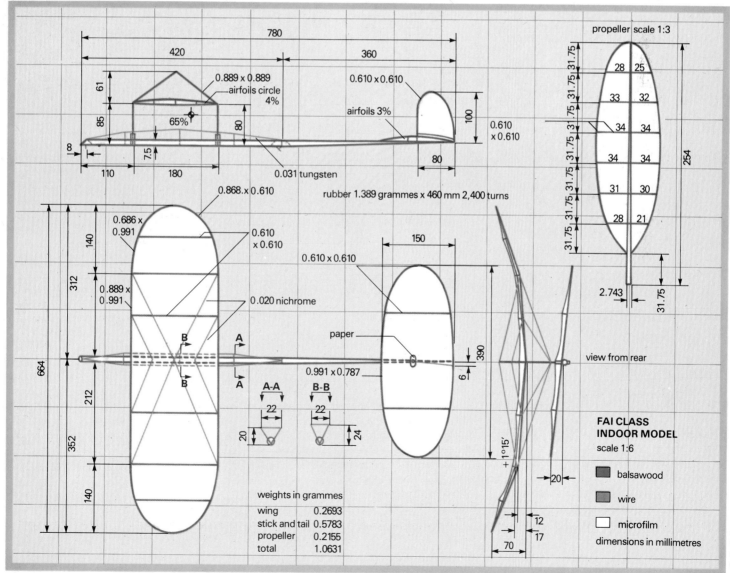

The preparation for any first flight is an exhilarating climax to many hours of craftsmanship and dedicated attention to detail. Holding the fuselage and propeller *(left)*, the modeller launches the frail, insect-like craft into the air and waits expectantly. If all goes to plan, the propeller paddles slowly *(below)* as the model climbs away.

final scores, so many of the 'hard luck stories' can be discounted.

Typically, the flight of a microfilm model can be divided into three stages—the climb under power, cruise at height and descent under low power—each of about the same duration. Models are hand launched—no undercarriages are fitted. For the uninitiated, the first surprising observation is the slowness of flight. The propeller turns at only about one revolution a second compared with three to four thousand rpm for full-scale aircraft and up to five times as fast for models powered by a combustion engine.

Critical stage

When the model is released, it paddles slowly away as it ascends. These first seconds of flight are critical: too much power from the rubber motor and the model will rear up and stall; too little power and the model will not reach sufficient height to make use of all the turns on the motor during the descent. Using the correct length of motor, number of turns applied and grade of rubber are also of the utmost importance. The quality of rubber varies from year to year, so the best vintages are carefully stored. Some modellers bury the rubber to exclude light and air, thus preserving it for many years. Such dedication to details is essential because the rubber motor must be wound to within a few turns of its breaking point; and, if snapped, the model is likely to be irreparably damaged.

The extended flight times of microfilm aircraft and the limited duration of ideal conditions result in a crowded 'sky' during competitions. And the insect-like qualities of these will-o'-the-wisp machines appear to extend to a natural attraction to one another, leading to frequent mid-air collisions. When two of these fragile, insubstantial models become locked together, they descend slowly to the ground in a sort of macabre mating dance like a pair of dragonflies.

Left to their own devices, the models would frequently aim for the walls of the building, with the risk of damage or being trapped by one of the structural members. To avoid this catastrophe, the rules of the competition allow for the guiding of the model away from an impending collision. Hydrogen-filled meteorological balloons are tethered to lengths of cord and placed in front of the flight path of the model to deflect it into a safer direction.

This gentle operation is not without risk. If the balloon rises to the steelwork, it might implode and disintegrate the model through the subsequent shock-wave. And an over-climbing model can become hung-up or trapped in the roof trusses, causing the flight to be declared terminated.

Less obvious dangers can catch out the unwary competitor. In one instance, a pigeon flew into the carrying box of an unfortunate enthusiast, destroying his models, but a mere fly-past is sufficient to wreak havoc with the flight pattern.

To see microfilm models lazily circling a giant airship shed is, on reflection, a fitting but saddening observation. Enthusiasts depend almost exclusively on such buildings for their competitions: if they disappear, as have the unwieldy inhabitants for whom they were originally built, a similar fate may well befall the seemingly unwieldy but delicate microfilm aircraft.

Military Technology: Electronic Warfare

Battle of the airwaves

With each of the superpowers having the capability of destroying the whole planet several times over, the emphasis in modern warfare has switched from weapons development to intelligence and counter-intelligence. This has led to the development of a whole new branch of military technology—electronic warfare—where the microchip is mightier than the missile.

Electronic warfare (EW) is normally broken down into three sub-sections: electronic counter-measures (ECM); electronic counter-counter-measures (ECCM); and electronic support measures (ESM).

Improved technology has made EW an increasingly complex area of military operations. ECM includes techniques which disrupt or interfere with radar systems as well as radio. ECCM are those measures which are taken to protect a radio or radar source: ECCM not only minimizes detection but, in effect, 'counter-attacks' the electronics of enemy weapons homing in on a radio or radar source.

ESM are the activities and equipment used for ECM and ECCM in a passive role. An example would be the warning equipment fitted to an aircraft to detect hostile radar emissions from a pursuing fighter or anti-aircraft missile.

ECM can be either active or passive. The simplest active form of ECM is to tune to the radio frequency employed by the enemy and, using a more powerful radio transmitter, blot out their weaker signals with 'mush' or electronic noise. The counter to this is to re-tune the radio, attempting as far as possible not to give the enemy any idea of how effective his measures have been. This is normally done by changing frequencies at a fixed time of day, but it can also be done on receipt of a code word from the central station on the radio net.

An example of passive ECM employed by Soviet forces is ground radar reflectors. These are multi-facetted metal shapes which can be hung in trees or placed on posts, and which give misleading radar echoes on the NATO radar used to monitor battlefield movement.

The most common passive ECM system, however, is *chaff* or *window*, developed dur-

Above Electronic warfare and gathering of intelligence goes on all the time. Spy aircraft like this Soviet Tu-20 Bear bomber (and its NATO equivalents) patrol foreign air-space, full of electronic equipment *(right)*, to compile data on radio and radar technology.

ing World War 2 and still effective. Chaff consists of thin strips of aluminium foil, or metallized glass or nylon fibres cut to match the wavelength of the incoming radar waves. They can be dispersed from a pod mounted on an aircraft or launched from a ship by means of a rocket.

An effective shielding system is essential for all but the smallest warships since they are a large target—ideal for tactical SSMs (surface-to-surface missiles). The British *Corvus* launcher can give cover against three missile attacks from a maximum of 16 launchers. The system is linked to a passive detection device which alerts the operators when the ship has been 'spotted' by the radar of an incoming missile.

Infra-red homing missiles can also be decoyed by flares with a very strong heat source. However, like the chaff system, infra-red ECMs have to be launched before the incoming missile has acquired a homing lock on its target. Too early, and the flare or the chaff will have dispersed, too late and the missile will have found its target.

Perhaps the best defence is to attack the radar of the attacking ship before it launches its missile. If it can be jammed or misled it will be unable to target its missiles.

Attacking ships, however, like fast patrol boats, now carry missiles which can discriminate between chaff and flares and the ships themselves launch their missiles from over the horizon so that the target vessel cannot get a radar picture of them or make visual contact.

Air forces are second only to navies as major employers of ECM measures. It is now standard practice to fit ECM pods to aircraft to protect them during operational missions.

Right Grumman's Intruder and Prowler *(inset)* were designed as ECM support aircraft for the US Navy. They can carry the ALQ-99 or APQ-92 jamming systems along with a variety of underwing pods. These are tailored to specific missions and can be used to jam enemy communications nets (air and ground), radar sets, and missile guidance systems.

Left ECM and ECCM are even more important at sea. Here attack (on an enemy's radar system) is considered the best defence against surface-to-surface missiles.

Right Radar can be fooled or jammed by using both active and passive ECM measures. The simplest counter is to use radar sparingly on other frequencies.

Military Technology: Electronic Warfare

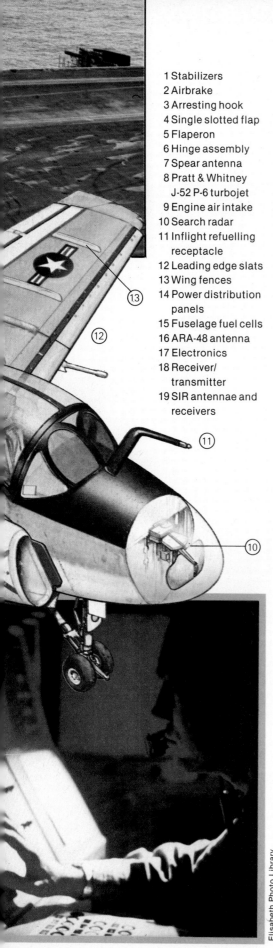

1. Stabilizers
2. Airbrake
3. Arresting hook
4. Single slotted flap
5. Flaperon
6. Hinge assembly
7. Spear antenna
8. Pratt & Whitney J-52 P-6 turbojet
9. Engine air intake
10. Search radar
11. Inflight refuelling receptacle
12. Leading edge slats
13. Wing fences
14. Power distribution panels
15. Fuselage fuel cells
16. ARA-48 antenna
17. Electronics
18. Receiver/transmitter
19. SIR antennae and receivers

These pods can be modified easily and replaced if damaged without the aircraft being grounded for a major overhaul.

Most SAM (surface-to-air missiles) or AAM (air-to-air missile) systems use either radar, infra-red or electro-optical guidance. The most difficult of these systems to jam is the electro-optical, since it relies partially on a human operator.

The Vietnam and Yom Kippur wars saw the most extensive use of ECM equipment. In the Middle East, Israeli aircraft used the well proven tactic of flying low in an attempt to confuse Egyptian radar by making it pick up ground 'clutter'—misleading radar echoes bouncing back off the ground. However, this put them within range of radar-controlled ZSU-23-4 anti-aircraft guns.

Air defence

Furthermore, when they climbed away they came within range of the SAM-7 *Grail*, a one-man anti-aircraft missile with infra-red homing. If they climbed higher to go beyond the 3,700 m range of the SAM-7 they were well within range of more sophisticated systems like SAM-2, SAM-3 and SAM-6 which had their own ground radar to track the enemy aircraft and guide the missile.

The Israelis had the benefit of American experience in Vietnam when it came to countering the SAM-2, but the SAM-6 was new equipment and caused considerable losses. Part of the problem is that SAM-6 uses semi-active radar guidance in which the missile homes on radar signals reflected from the target by a tracking radar on the ground.

By the end of the 18-day war the aircraft losses had fallen dramatically, however, and though the Israelis and Americans may have developed some type of ECM protection against the *Straight Flush* radar of the SAM-6 it may have been that the Egyptians had simply run out of missiles.

The Vietnam war was also a major proving ground for airborne ECM equipment and many of its developments are still in service with improved and updated electronics. Escort aircraft like the Douglas EA-3A and EA-3B began life as tactical bombers but were converted to take electronic equipment and operators in the bomb bay and fuselage.

Their role was to escort bombers on missions into North Vietnamese air-space and jam frequencies used by enemy radar and radio. They would orbit outside the range of the defences and use equipment which could either swamp enemy signals with a far stronger signal—*noise jamming*—or use the same frequency but modify it to give the enemy receiver a false picture or sound—*deception jamming*.

Noise jammers generally sweep the waveband, locate a frequency in use and then concentrate their energy there. But they can also be used over a wide band range as *barrage jammers* to completely jam a waveband. Deception jamming is a more economical system but it requires prior knowledge of the enemy frequencies and equipment.

Chaff is also used by escort aircraft, and can be laid in long swathes to screen bombers, or as a false signal to give enemy radar the impression that an attack is being directed at another target. A mix of chaff, false radio traffic and a few bomber attacks can give enemy ground controllers the wrong picture and make them direct their efforts away from the real target area.

The aptly named Grumman Intruder and Prowler aircraft carry a vast array of ECM equipment. The Intruder has over 30 different antennae to detect, locate, classify, record and jam enemy radar while the Prowler carries five separately powered pods with ten jamming transmitters. These aircraft could be used most effectively against an enemy to jam all military and commercial transmissions. Unable to contact the world by radio and telex, and with no television or radio serviceable in their country, the enemy would be temporarily crippled.

Electronic smokescreens

Ground troops have the least ECM equipment. However, laser-guided rockets and shells are entering service, and infantry vehicles such as tanks may soon be equipped with detection devices to show if they have been 'designated' by a laser as a potential target. Counter-measures could well include powerful lasers to blind the incoming missile or hot flares to confuse missiles using infra-red homing sensors.

For the soldier on the ground, hand-held passive detectors have been introduced which can tell him if he has been located by a battlefield radar, while passive night sights can detect an enemy using infra-red equipment to see in the dark. Smoke and chaff can also be used by infantry to confuse optical and radar-controlled missiles.

Electronic counter-counter-measures fall into two groups: those measures that will defeat enemy ECM and those aimed at his ESM. The basic ECCM technique is to devise radio equipment which is able to switch frequencies very quickly as soon as it is jammed.

2231

Military Technology: Electronic Warfare

Top Radar-guided missiles are a potent threat to aircraft. But US strike aircraft now carry anti-radiation missiles *(above)* that home in on an enemy's radar signals and fly down the beam to the transmitters *(right)*.

Left Even Fylingdales' early-warning station is not invulnerable to ECM. Jamming a ground-forces net *(above)* may be unnecessary—careless talk can tell the enemy a great deal.

An alternative is to make equipment which can receive or transmit across a wide band so that though ECM may be able to jam on one frequency it cannot jam the others. If the equipment is designed to both switch quickly and operate on a wide band it becomes even harder to jam.

This is the principle of a new British-designed combat radio system, Racal's JAGUAR V (or Jamming and Guarded Radio VHF). This device changes its transmitting frequency several times a second in a random sequence governed by a code selected by the operator. At the receiver an identical random sequence generator alters the receiver frequency to conform with the transmitter's. To ensure that the two are synchronized the transmitting set sends out a 'synch' signal at regular intervals. The receivers on the network pick this up and automatically change frequencies in time with the transmitter. This makes the network virtually impossible to jam or intercept, and difficult to locate.

One of the more dramatic duels between ECM and ECCM is the use of radar-homing missiles. The Americans pioneered this tactic in North Vietnam with aircraft code-named Wild Weasel. These aircraft carried *Shrike* or *Standard* anti-radiation missiles (ARM) and deliberately flew to enemy SAM sites.

Anti-radiation missiles are designed to fly down a radar beam to its source and so destroy either the tracking or guidance radar. Aircraft armed with ARMs carry advanced ECM equipment to warn them when they are being tracked by radar.

At present ARMs are too expensive for widespread use; but, as advances in microprocessor design reduce the cost of the advanced electronics these missiles contain, ARMs will probably become part of every aircraft's protection. This will allow one aircraft out of a group to engage a target without all the missiles being launched at the first enemy radar that is switched on.

The most important part of any ECM or ECCM system is intelligence on enemy radars, radios and electronic equipment. This is derived from the electronic support measures of aircraft and ships. The basic ESM is a passive receiver which detects emissions over a wide frequency band. It can identify radar by its pulse width, repetition, frequencies and scan rates.

For more exact information tunable receivers are linked to minicomputers or microprocessors. This type of equipment is fitted to the Soviet 'trawlers' that follow NATO exercises and the Soviet bombers that shadow fleets at sea or attempt to penetrate NATO air-space. So sophisticated are some types of ESM that NATO troops training with anti-aircraft missiles well away from the coastline are kept constantly informed about Soviet activities so that their own radar cannot be monitored.

Advanced ESM equipment can work in very cluttered electromagnetic environments and, using a 'library' of recorded frequencies, identify new emitters in a particular area. They are an essential part of ECM since in a conflict they could jam the most threatening radar systems or radio sets.

Monitoring

Any system is only as good as its operator—and even the two year draft for Soviet soldiers may become too short a time to train a man to become a good electronics technician. The West has attempted to resolve this problem by developing simulators which give radio and radar operators the feel of working through jamming. Since these simulators are closed systems they produce no emissions which can be detected by enemy ESM equipment.

It is essential also that operators do not let the enemy recognize them by some individual quirk or accent. With a good ESM library an enemy monitoring unit could quickly trace the movements of a force simply by listening to the way in which different radio operators speak. All operators are therefore trained to speak in as uniform and clear a way as possible. This makes them harder to tell apart and easier to hear during interference or jamming.

During a long war ECM operators can be trained to imitate enemy call signs and, by sending a small number of misleading messages, destroy the credibility of an entire enemy network. For this reason an unknown call sign coming on the air can authenticate itself by using a simple number code that is held by all the operators in the radio net. If the call sign cannot authenticate itself it is then ignored by the entire net.

Jamming remains, however, the easier option, particularly if the radio operators are using an esoteric code language. Indeed, during WW 2 the US 82nd Airborne Division used American Indians as radio operators, purely because no enemy would be able to understand them anyway!

Sports Technology: Tennis

Science serves an ace

Since tennis was first formulated as a leisurely recreation for the fashionable society of the 1880s it has seen many developments of racket, court and ball. But even more revolutionary than the changes of 'hardware' are the latest developments of 'software'—teaching methods that rely on biomedical analysis to computer-analyze performance. Such methods enable players to modify their game to resemble that of an ideal, computer-generated 'model' —or even to imitate a favourite tennis star. If the trend spreads the result may be thousands of players around the world whose styles of play bear an uncanny computer-trained likeness to the games of McEnroe, Connors or Borg!

Pioneers of the new technology of biomedical analysis are Vic Braden, psychologist and one of the best tennis teachers in the world, and Gideon Ariel, a former Olympian. At first glance, their Coto Research Center south of Los Angeles looks like an ordinary health club. But a closer look reveals that it is among the most advanced sports complexes in the world.

The researchers started by investigating how much work a player's heart has to do during a game of tennis. To do this they fitted lightweight radio sensors to the tennis players' chests that could transmit the heart rate data to a computer in the club-house. Here a cardiologist monitored what was happening during a match. Simultaneously, eight sonar sensing devices around the court fed the central computer with information about the players' motions, to indicate how hard they were working.

Braden is now fitting athletes with special headgear which contains a complex set of lenses. The device is fitted with a light source that bounces light off the player's eyes and into a camera at the back of the helmet. Since the camera commands the same field of vision as the player, a bright spot of light on the developed film will later indicate exactly where the player's eyes were focused. This information helps the teacher to understand the player's style of play, and enables him to suggest improvements.

Braden and Ariel also plan to monitor top athletes under the stress of competition by taking high-speed cameras to track meetings, tennis tournaments, basketball games, and other athletic events. Braden is interested in knowing what happens when the athlete is performing 'live' so that he can compare figures with those obtained in the lab.

Filming is the first of a series of carefully co-ordinated steps in the analytical process at Coto. After the film is developed, it is scan-

Analysis of a player's level of exertion, plus studying eye movements using a special lensed mask *(centre)*, raises standards a step closer to Borg *(far left)*, Connors *(left centre)* and McEnroe *(below right)*.

ned and fed into a computer-digitizer. This process produces simplified stick figures which enable Ariel to study an athlete's motions in a three-dimensional sequence on the computer screen. Ariel then compares this motion analysis with that of an 'ideal' model motion stored in the computer's memory. Then by comparing the tennis strokes of player and model Ariel can suggest corrections. Thus biomechanical analysis by computer enables Ariel to instruct the player on how to achieve the best results.

Although it is becoming increasingly popular, Braden and Ariel's work has so far only benefited a handful of the tennis elite. Of more immediate relevance to the rest of the tennis world has been the rapid improvement in tennis racket technology. After all, even the best players can only utilize their skills fully if they have the right racket for the task. The racket is the link between the player and the ball. It must have exactly the right amount of hardness and flexibility, and its strings must have the right tension to give the ball a good 'throw'. But there must be no sacrifice of 'feel' or control. Finally, the *sweet spot,* or effective hitting area, must be as large as possible to give the player the maximum margin for error, yet the frame must be within the limits set for maximum dimensions and weight.

Until the 1960s all rackets were made of one piece of laminated timber. Wood is a natural composite material consisting of

Sports Technology: Tennis

fibres of cellulose bonded with *lignin*—a natural adhesive. However, the exact structure of wood varies from sample to sample, and also depends on the type of tree the wood comes from. Dozens of types of wood are used in making rackets, ranging from common ones like ash, beech, maple, hickory and mahogany to such exotic timber as sycamores, bamboo and obeche. Each wood has its own particular blend of qualities: maple is the stiffest and hardest, bamboo is the most flexible, and the others have properties that lie somewhere in between. The skilful combination of these timbers in a multi-laminated frame gives the racket the essential quality of 'feel'.

Now, however, the introduction of artificial composite material is replacing wood and has marked a revolution in tennis technology. The composite materials used are similar to wood, consisting of long strong fibres embedded in a bonding matrix. But the artificial materials are much stronger than wood, more predictable in their behaviour and just as light.

An example of the composite racket is the Head Ashe Competition which has an extremely strong and lightweight foam core combined with glass-fibre laminate, and surrounded by an outer skin of aluminium alloy for strength and rigidity. The Max 150 G, on the other hand, has a core of carbon fibre-reinforced nylon filled with low-density polyurethane foam. Medium-density foam is concentrated in the shaft and handle to give good balance and feel.

The graphite used in composite frames is pure crystalline carbon made by drawing out organic carbon-based fibres through low and high temperature furnaces surrounded by in-

Above left to right Wilson T 2000 with unique trampoline wire stringing; AMF Head Boron; Prince 'Jumbo size' racket with extra large 'sweet spot'; Phillips-Moore—Canadian snow shoe shape with long head; Fischer Superform—diamond shape; Donnay Borg pro racket—long gripped.

Left The Head Arthur Ashe Cup. It contains additional glass-fibre laminates in the head to increase resistance to distortion and enable it to play faster shots.

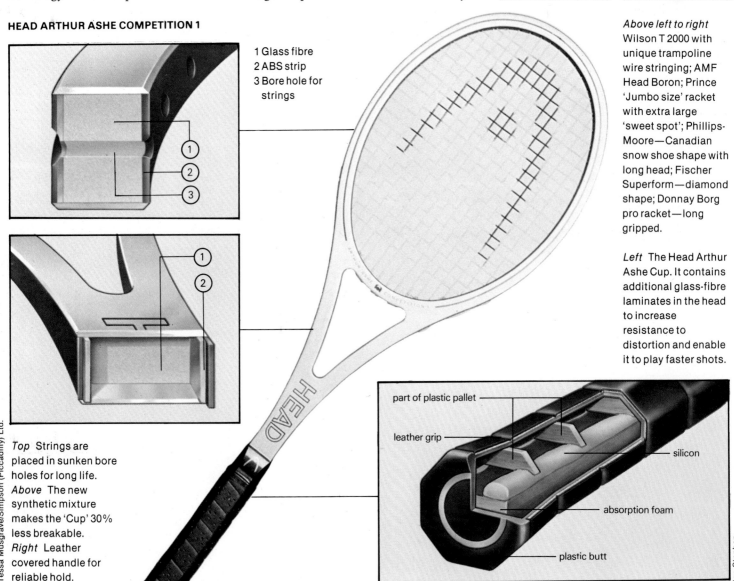

HEAD ARTHUR ASHE COMPETITION 1

1 Glass fibre
2 ABS strip
3 Bore hole for strings

Top Strings are placed in sunken bore holes for long life. *Above* The new synthetic mixture makes the 'Cup' 30% less breakable. *Right* Leather covered handle for reliable hold.

DUNLOP MAXPLY 150G

ert gases. At the second stage of heating all the molecules are burned off except for the pure carbon which remains as stiff light fibres whose molecules are all aligned in one direction. By aligning the carbon molecules in this way the strength of the material can be made twice that of steel, but with only a quarter of the weight. Finally, carbon fibres are combined with epoxy resin to produce the material used in frame making.

Glass fibre is also used in racket construction. It consists of thin, strong glass fibres embedded in a matrix of epoxy resin or plastic. The result is a material that can be tailor-made to have the right characteristics for a particular task. The tensile strength (resistance to stretching forces) of glass fibre is equal to steel, and it is also extremely resistant to twisting of the racket head.

Light and durable

Boron is another new material first used for tennis rackets in the Head Competition II. Boron is chemically related to aluminium, and like aluminium it is light and stiff, but it is also much harder than aluminium (and considerably more expensive). Boron fibres are produced by depositing vaporized boron onto fine tungsten wires. The resulting fibres are then embedded into a matrix to produce a material of very high tensile strength and low weight. These properties enable boron to be used as an additional lamination of high durability and resistance.

Both boron and graphite are now combined with a variety of other materials such as

Right The Max 150G gives resilience, accuracy and power by mixing graphite fibres and nylon. The stiff design also helps prevent elbow injury problems.

- stringing tension 58 lb
- gut stringing
- hard epoxy paint and synthetic matt lacquer for finish
- leather grip

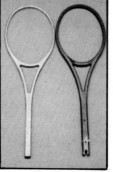

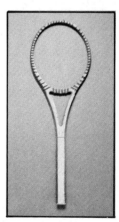

Centre left A composite material with nylon and graphite fibres is prepared for moulding.
Left The graphite composite is injected around a metal core to the desired shape.
Above left Internal pillars are formed around each string hole to reinforce the structure.
Above right The frame is then filled with foam of varying density for balance and damping.

Sports Technology: Tennis

wood, nylon, aluminium, glass fibre, and these composite frames are now taking an increasingly large share of the world market.

Before synthetics came into use great advances had already been made with metal racket frames. There had been efforts to produce steel frames in the 1930s—even to the extent of using wire to string them. But there was no production model available until the Lacoste T2000 model with trampoline-type stringing was launched in France in the late 1960s.

The Lacoste T2000 racket led to a whole family of Wilson T model rackets which in turn spawned a variety of steel and metal rackets. The unique 'trampoline' wire stringing method of the T2000, 3000, and 4000 rackets tended to fling the ball, giving added power but less control—a combination of characteristics which had to be used with skill. The T3000 also included a throat brace to stiffen the shaft, and the T4000 had a tungsten wire located inside a circular plastic ball at the end of the grip to act as a damper to reduce vibration.

Introduced a little later was the T5000 which used an alternative type of damper (also patented by René Lacoste) and an entirely new 'Increased Density' string pattern to give a greater density of strings in the primary hitting area or sweet spot.

Steel loses ground

But although it has been quite widely used, steel has steadily lost ground to aluminium which is becoming the most accepted metal for rackets. Aluminium is more durable and less flexible than steel, but offers about the same degree of control as wood.

In its pure state aluminium is not particularly strong, but when alloyed with other substances such as magnesium, copper and silicon it has a high degree of tensile strength. It is also light, corrosion-resistant and easily machined or worked.

The development of aluminium has enabled Howard Head to develop a large size Jumbo racket with an increased effective hitting area. Head developed the racket to overcome the tendency of a racket to twist when a ball is hit off centre.

Head's first attempt to solve the problem was to weight the sides of the racket frame, but he found that this often resulted in the racket breaking. He then turned to the idea of using a large racket since he knew that additional width had the effect of counteracting the turning effect of mis-hitting. This is because turning resistance is proportional to the square of the extra width, so that widening the racket by 20 per cent increases the twist resistance by 40 per cent. The result was a patented design for a racket that was wider, and some 7 cm (3 in.) longer than conventional designs.

In July 1981 the International Tennis Federation adopted a proposal which limited the size of rackets to 81.28 cm (32 in.) in overall length and 31.75 cm (12½ in.) in overall width, with the strung surface not to exceed 39.37 cm (15½ in.) by 29.21 cm (11½ in.). The Prince 'Classic' which Howard Head developed comes just within these parameters, as do the models which have followed in the same Jumbo size—the 'Pro', 'Woodie' and 'Graphite'.

There are abstruse mathemetical formulae which account for the improved performance of these rackets, but in simple terms it is due

ARTIFICIAL PLAYING SURFACES

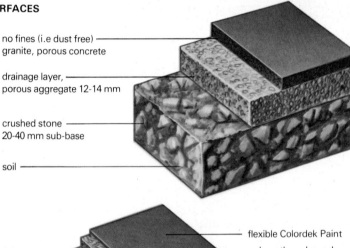

Right Pupils learn faster on Tennisquick because the surface gives a consistent bounce at all times. So the emphasis is on learning rather than on coping with uneven surfaces.

- no fines (i.e dust free) granite, porous concrete
- drainage layer, porous aggregate 12-14 mm
- crushed stone 20-40 mm sub-base
- soil

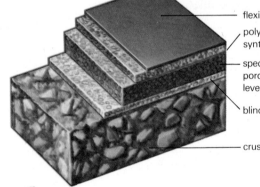

Right Tenniprene is porous so it can be played on after rain stops. Snow can be brushed off and, as the surface is a good thermal insulator, ice melts as soon as temperatures start to rise after a freeze.

- flexible Colordek Paint
- polyurethane bound synthetic rubber granules
- specially formulated, porous bituminous levelling course
- blinding stone
- crushed stone foundation

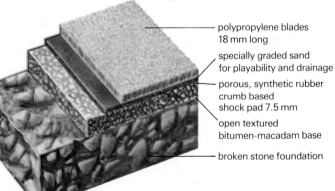

- polypropylene blades 18 mm long
- specially graded sand for playability and drainage
- porous, synthetic rubber crumb based shock pad 7.5 mm
- open textured bitumen-macadam base
- broken stone foundation

Left Sporturf is a bladed granular surface that adopts the principles of biomechanics to give an all-weather pitch that plays like grass, not a carpet. Unlike grass it does not wear thin or get spongy after rain.

Right The 'Tennis Partner' returns balls at precisely the right speed, height, and position no matter how hard each is hit. *Below right* The 'lobster' throws out balls at 2, 4 or 8 second intervals. *Below* When it is 'out', a ball triggers the photoelectric cell of the automatic line judge. *Left* Flushing Meadows, USA, home of the US Open tournament.

to the sweet spot (primary hitting area) being roughly 40 per cent larger than normal—because the Jumbo offers more effective hitting area it also provides a greater margin for player error.

Most manufacturers have now introduced models similar to the Prince, and these have become known as 'mid-size' tennis rackets. Their stringing areas are between 18 per cent and 30 per cent above the standard racket size. Mid-size rackets have greater manoeuvrability and may well become established as the ideal size.

An interesting recent development in racket design has been made by Fischer of Austria, one of several ski makers to put their technological know-how developed in ski manufacture to use in producing tennis rackets. The Fischer 'Superform' shape has an elongated sweet spot in a standard model racket shape.

An even more bizarre new design is the Phillips-Moore racket which resembles a Canadian snow shoe in shape—with its very long head and short handle. The shape is designed to give better control. Another unusual design is the 'hold-less' racket developed by G.W. Anderson and Dr C.N. Reid. The frame is made from aluminium alloy and the design eliminates traditional transverse stringing holes by substituting a circular strip around the inner edge of the frame which can be milled away at the appropriate places so that eyelets can be produced through which to thread the strings. This design avoids the necessity to make holes in the frame which might act as stress concentration points, starting fatigue cracks and weakening the frame.

As well as developments in tennis rackets, there have also been developments in other equipment for the game—such as court surfaces. Red shale is the surface upon which the major Continental tournaments and Junior Wimbledon are played. It is a good playing surface but, like grass, it is expensive to maintain. The advantages of the loose, crushed brick dressing are that it allows a player to slide into a shot, while the mark left by a ball-bounce helps the line judge decide whether a ball is in or out.

Constructed of water-bound materials including brick and granite, shale courts are porous and, if correctly maintained, enjoy a life of at least 20 years. But on a hot day a shale court needs regular watering and brushing, and the shale is vulnerable to winter frosts and snow. The cost of a new court, after levelling, is comparatively low, but it is a costly investment in the long term since it requires constant maintenance.

The type of court that is gaining popularity fastest is the hard, all-weather, non-attention variety. The advantages of this type are that once laid, they require little or no maintenance for up to 15 years. The all-weather courts are composed of up to four layers, with each ingredient scientifically designed to provide uniformity of bounce together with weather resistance and comfort for the user.

One of the most interesting recent developments is the plastic mesh surface that can be laid over existing courts. These non-bonded courts can be laid and removed with ease, and their qualities are remarkable. The bounce is true and lively; the feel underfoot better than almost anything except grass, and the upkeep is minimal.

Synthetic courts

In an attempt to reproduce the more desirable qualities of a grass court, synthetic courts that look and play like turf are now being designed. Omnicourt, for example, a combination of polypropylene fibre and blended silicas, has been tried at the All-England Club, Wimbledon.

A recommended tournament surface is Supreme, a four-ply sandwich construction with a fibrous face coat and vinyl foam cushion backing. Supreme is renowned for its fairly slow pace, although this can vary according to the surface it is laid on. Other carpets, such as Nygrass and Bolltex, are much faster, while En-tout-cas's Tennilux, a porous carpet of needle-punched polypropylene fibres that can be laid outdoors, is also quite a slow playing surface.

Another innovation in this fast-moving sport is the 'Tennis Partner' practice device from Sweden. This device will automatically return balls to the player at precisely the right speed, height and position. It is an ingenious piece of equipment that can be assembled in minutes, on any surface. It uses a PVC 'half-court' that can be set up to a pre-determined angle for ground strokes, service, or volley practice.

Technological tennis aids such as the ingenious 'Tennis Partner', the latest range of large-size composite rackets, and instruction with computer-based biomedical analysis are transforming tennis. The result will be better play, not just at the top of the league but at all levels of the game.

Machine Technology: Civil Engineering

Road building: the way ahead

Roads are one of the few Man-made features visible from space. Vast highways, like Martian canals, cut through forests, passing over mountains, across deserts and spanning rivers and valleys as they stretch from horizon to horizon. Yet this endless network which enmeshes the Earth is the product of little over 60 years' endeavour and represents the application of tremendous engineering skills—dedicated to the internal combustion engine and a vision of freedom.

The design and construction of a modern road is an outstanding engineering feat requiring many years of planning and the application of a vast range of scientific and technical disciplines. Once the decision to build has been taken, one of the first tasks is an exhaustive technical survey that yields a wealth of essential information. Using aerial photographs taken from high-altitude aircraft or even satellites, accurate profiles and cross-sections of the terrain are produced. These give an idea of the many problems engineers will meet, as well as specific information about soil characteristics and geological formations. Computers may also be used to calculate the amount of manpower and machinery needed to complete the project on schedule and to give a detailed analysis of the various costs.

The new road

In planning the route, the engineers make projections of traffic volumes 20 years hence and try to plan for the needs of industries and communities that as yet do not exist, but will eventually depend on or be affected by the new road. Practical considerations, such as the amount of excavations and in-filling along each section of route, the building of drainage and damage to agricultural land, are essential but equally important are the likely effects of the new highway upon population centres, and the level of noise and pollution from exhaust emissions.

Also, the line of the highway should blend naturally with the topography of the land in order to keep construction simple and minimize environmental impact. Long-distance driving against the natural 'grain' of a landscape can greatly increase stress and lead to driver fatigue. Allowances have also to be made for other road systems, waterways and game trails that cross the line of the new road. To ensure nothing is overlooked, designers and engineers work closely with other technologists, from city planners to sociologists, and make personal examinations of the route and its terrain.

The load-bearing qualities of a road depend crucially on its foundation, which consists of concrete (right) laid on a compacted earth base (below). The finished road (centre) blends in with the lie of the land but the road is entrenched slightly so that noise and emissions are contained and the scenery is little disturbed. The latest trend is to plant the central reservation with trees to provide a light screen.

In the USA, engineers planning one stretch of the US Interstate Highway System walked the entire route from Albany, the state administrative centre, to the Canadian border some 290 km (180 miles) away. As a result, the road they planned and built, Interstate 87, has split-level roadways on each side of wide, landscaped park strips planted with trees. The road blends unobtrusively into the natural rise and fall of the land, and has rest areas and scenic views for travellers, as well as underpasses for wild animals.

The path of the road was drawn, using a computer, as a series of contours to give an exact vertical profile along the road's centre line. The *grade line* is then established, showing the proposed level of the road throughout its length. The grade line is used to calculate the amount of earth-moving necessary, and is adjusted to establish the ideal balance between cuts and fills.

As early as the 18th century, it was realized that the foundation, rather than the surface, is a road's load-bearing structure, but the design of roads really only developed to keep pace with increased weight and traffic.

The modern expressway carries an almost

continuous traffic of heavy vehicles travelling at high speeds. The weight of commercial loads increases as fast as local and national regulations will allow, for the larger the load, the greater the economic advantage to the carrier. But to reduce the damage to roads and the buildings alongside them, the maximum load on each axle is restricted because it is the axle load that is significant, rather than the overall weight. One way to increase overall weight without exceeding

Usually, trunk roads intersect at roundabouts or light-controlled junctions; but to maintain traffic flow at motorway intersections *(above)*, a system of underpasses and overpasses is preferred—despite the large space needed.

axle load is to increase the number of axles to distribute the load. In any event, the combined effect of great numbers of high-speed cars and slower but damagingly heavy trucks is a need for roads with massive foundations.

The path of a new road is prepared by clearing the land. Vegetation is removed by bulldozers and the earth levelled by giant scrapers, some of which can remove 50-60 tonnes of earth at one go. For a sound, stable foundation, the earth base and any in-fill material is compacted mechanically to avoid subsequent movement or subsidence due to settlement. One compactor consists of rollers vibrating 3,400 times a minute.

To evaluate the amount by which soil or other material can be compressed on-site, an analysis is made of their molecular and granular structure and of the amount of moisture they contain. This information is sometimes supplemented by data from nuclear probes. These are used to emit radiation into the ground and measure the amount that is scattered back to the instrument—which becomes less as the earth is compacted. The instruments give readings of moisture content and soil density.

Left Reinforced concrete is among the cheapest road-building materials.

Where embankments or earth-fills are to be built across wet, unstable ground, vertical sand drains are constructed on the site to increase stability. Sand is poured into boreholes drilled into the base material. A layer of sand or gravel is then placed over the sand-filled drain holes and the fill added on top. The weight of the fill forces water from the wet material up the sand columns, to be drained off through the sand or gravel layer.

After compaction and drainage, the road is ready for the pavement—the part of the road above the original earth base. Pavements are built up as a base, sub-base, and surfacing.

Flexible pavements

Modern pavements can be flexible or rigid. Flexible pavements have a base and sub-base of natural aggregates, such as gravel or sand, topped with a surfacing of aggregate and asphalt or some other bituminous substance. Alternatively, the base and sub-base might comprise soil, sand and gravel, to which stabilizing materials, such as Portland cement, coal tars, lime or various chemicals are added to make them more resistant to moisture increase and weakening.

There are four main surfacing processes for flexible pavements. Their application is suited to the volume of traffic the road is expected to carry. Usually, aggregates are mixed with a cementing agent in the roadway itself and laid to give a strong, waterproof surface. But where the volume of traffic is high and the climate severe, the aggregates are best mixed in a central plant.

As well as being waterproof, aggregate-based roads have good riding qualities. High-quality aggregates of various sizes ranging from dust up to walnut size are heat dried, mixed at a high temperature with asphalt, cement or semi-solid tar and transported to the roadside. It must then be well rolled before it cools. Plant mix surfaces have good frictional resistance and are easily improved by the addition of further layers.

Rigid pavement is used most commonly on heavy-duty highways. In this type, a concrete surface of Portland cement is applied directly to the compacted earth. This layer might be anything between 150 and 300 mm (6-12 in.) thick according to the sort of loads it is expected to bear. Batches of concrete and aggregate mix are prepared at a central site

MAIN FEATURES OF THE BK95 PAVER

1. Rollers enable contact with tipper wheels
2. Hopper with 13-tonne capacity—sufficient for continuous paving between refills
3. Conveyors—with automatic feed control and infinitely variable speed—move materials from hopper to rear. The flow gates are independently controlled and two wear-resistant augers distribute material across the screed width
4. Solid-tyred steerable wheels
5. Pavement height adjustment
6. Transmission drives the road wheels through a gearbox, roller chains and a differential to give infinitely variable working speeds
7. Variable-width screed
8. Diesel- or gas-fired heaters
9. Hydraulic hoses
10. Gear lever
11. Heater igniter
12. Foot brakes
13. Throttle
14. Conveyor speed control
15. Movable console
16. Six-cylinder diesel engine
17. Flow gate gauge

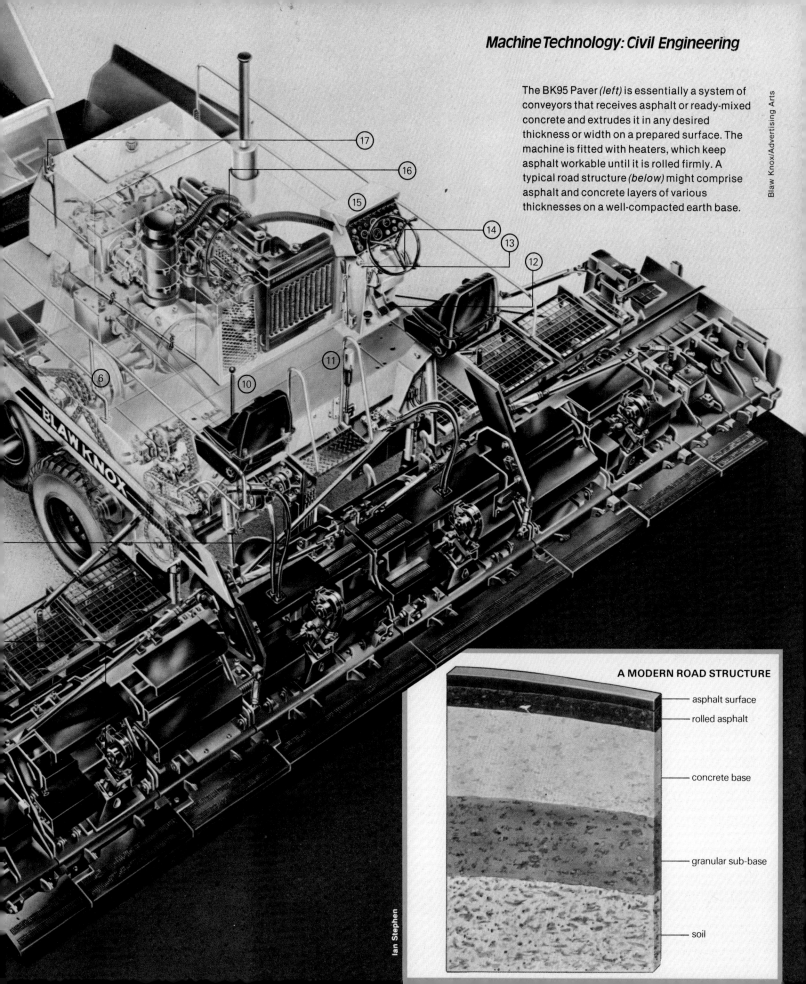

Machine Technology: Civil Engineering

The BK95 Paver *(left)* is essentially a system of conveyors that receives asphalt or ready-mixed concrete and extrudes it in any desired thickness or width on a prepared surface. The machine is fitted with heaters, which keep asphalt workable until it is rolled firmly. A typical road structure *(below)* might comprise asphalt and concrete layers of various thicknesses on a well-compacted earth base.

A MODERN ROAD STRUCTURE
- asphalt surface
- rolled asphalt
- concrete base
- granular sub-base
- soil

The construction of a new road begins with clearance of the route which might involve the felling of trees and demolition of houses. The route is then levelled by excavation or in-filling (1) to form a sound base. This is compacted (2), cleared of excess soil (3) and rolled (4). Any irregularities are graded (5) to ensure an even surface and structures such as bridges (6) and underpasses—which are valuable for wild animals—are constructed. The foundation —of varying thickness depending on the flow and weight of traffic—is then applied (7) to the earth base, overlaid with asphalt (8) and rolled smooth. Finishings are applied—including guard rails (9), reflecting studs and white lines (10). At all stages, surveys *(bottom right)* are essential to maintain direction and dimensions of the road. Usually, the route would be surveyed so that the planners can assess its suitability for present and long-term needs even before building is confirmed.

and trucked to the roadworks where they are fed in front of a paving 'train'. Because of the shrinkage that occurs in hardening concrete, there is a tendency for it to pull across the layer beneath, causing a build-up of stresses that lead to cracking in continuous concrete masses. Climate and moisture variations also affect the volume of the concrete once it has hardened. Rises in temperature or moisture content can make it swell, whereas a fall in temperature or moisture content causes shrinkage. On a continuous, unreinforced concrete motorway, direct sunshine can heat the surface of a 200 mm concrete slab to a temperature of 14°C above that of the bottom of the slab. The warping that accompanies these changes is a further cause of cracking. To counteract these tendencies, joints are often built into rigid pavements.

Jointed roads

Joints that compensate for contraction or expansion are essential in all long structures—from railway lines to bridges. In roads, grooves cut or moulded to a certain depth across the pavement ensure that cracks due to contraction occur in planned positions. Some joints extend through the entire concrete layer to compensate for expansion. The gaps between the segments are sealed against water and dust, and sliding steel sections across the joints provide an even surface. Longitudinal tied joints are used at the edges of highway lanes and are underpinned with steel tie bars to stop them opening up.

Building joints into long sections of road add greatly to the cost. For such projects, the concrete is reinforced with steel, enabling rapid construction and a jointless road. The pavement consists of a layer of concrete on which a mesh of reinforcing steel is overlaid.

The longitudinal steel members in the mesh withstand the tensions caused by temperature and moisture variations and hold cracks tightly together where they occur. Research into road building has resulted in an efficient, high-speed technique that employs a paving train. This is a mobile frame that straddles the road and moves slowly forwards on either tracks or wheels. Any number of stages can be built into the train, which can be used for laying both flexible and rigid roads. On a reinforced concrete structure, road trains can move forwards at a kilometre a day on the final layering stage.

Progress on a jointed road is slower. A typical train might consist of a spreader for the first course, supplied from the side by a tipper truck. This is followed in turn by a dowel bar placer and a longitudinal tie-bar placer to create the joints. A second spreader seals in the placed steel, and is followed by a second compactor and finisher to form longitudinal joints and finish the layers.

Some surfaces are brushed while still wet to give a non-skid texture, but the regular patterns of the brush can cause a high-pitched whine when driven over at certain speeds. The problem is solved by attaching a grooving machine to the train which makes irregular, anti-skid grooves in the road surface that are also noise-free.

Wealth of experience

Most major industrial nations are constantly involved with updating their highway systems and building new ones. Roadbuilding brought to light many technical and environmental problems but many lessons have been learned. In fact, there is a wealth of experience on which today's road designers can draw. These include the need to by-pass towns to avoid damage to buildings, surfaces that will remain non-slip in all weather conditions, central reservations to protect drivers from dazzling headlights, and consideration for wildlife, whose natural habitat might be affected. These and other experiences will help to ensure that tomorrow's roads don't lead to ruin.

Frontiers: Astronomy
The X-ray eyes

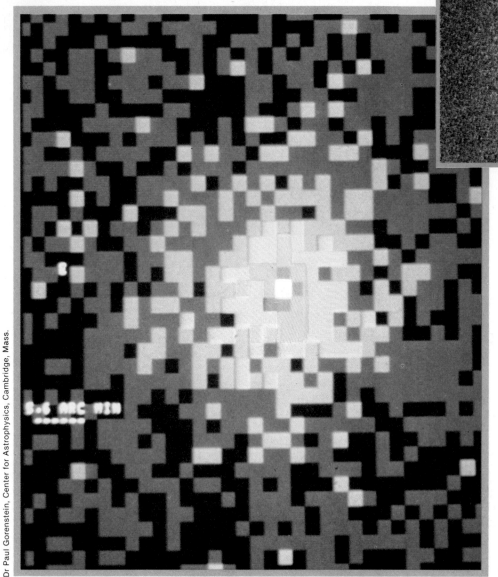

The radio galaxy M87, a giant galaxy which harbours a massive black hole, as seen by an X-ray telescope *(left)* and through an optical telescope *(above)*. X-ray observatories can pick up radiation from the swirling hot gas streams. The optical picture has been computer-coded to show different brightness levels, which makes the galaxy's famous jet stand out as an elongated blue streak.

Of all the radiations other than light which bombard the Earth day in, day out, X-rays have revealed the most dramatic pictures of the universe. X-ray telescopes show us pools of searingly hot gas millions of light years across and enveloping whole clusters of galaxies; nearer home, they reveal clouds of hot gas rushing away from old supernova explosions and the streamers of gas being drawn into a black hole. Yet X-rays cannot penetrate the Earth's atmosphere, requiring these tantalizing views of the universe to be snatched during brief rocket flights or from satellite observatories. All this is astronomy of the space age.

X-rays are electromagnetic radiation with wavelengths about a hundred times shorter than visible light, and are produced when a fast-moving electron is suddenly stopped. Terrestrial X-ray sources—such as hospital scanners—employ a beam of electrically accelerated electrons hitting a solid metal target. But out in the void of space, matter is rarely concentrated enough to provide either an adequate source of electrons or a barrier to their progress. Hence X-radiation is relatively scarce and where it does occur it indicates an area of great scientific interest.

Where gas is extremely hot (above 1 million°C), electrons are stripped from the atoms and collide with nuclei fast enough to generate X-rays. The spacing between nuclei in the tenuous gas of space is vast, but the gas clouds are so huge that any fast electron will eventually hit a nucleus and produce X-rays. The process is known as *thermal bremsstrahlung* ('thermal braking radiation').

It was long thought that no gas in the universe could be hot enough to generate X-rays—except in the nuclear reactions at the centres of stars, which are well hidden by their outer layers.

The Sun's corona

The surfaces of the hottest stars are around 40,000°C, while our own Sun's surface *(photosphere)* has a temperature of a mere 5,500°C. But in 1942, the Swedish physicist Bengt Edlén discovered that the Sun's outer atmosphere exceeds one million degrees. The spectrum of light from this region—the *corona*—had long been a puzzle, which Edlén resolved by interpreting it as the natural light emissions from very hot gas.

Such gas emits far more strongly at X-ray wavelengths; but these can only be detected from heights of 100 km (60 miles) or more above the bulk of the Earth's absorbing atmosphere. After World War 2 American scientists captured dozens of German V-2 rockets, and for the first time could launch X-ray detectors to these heights. A team from the US Naval Research Laboratory began searching in 1946, and finally detected thermal X-rays from the Sun's corona during a rocket flight in 1948.

Succeeding studies of the 'X-ray Sun' have

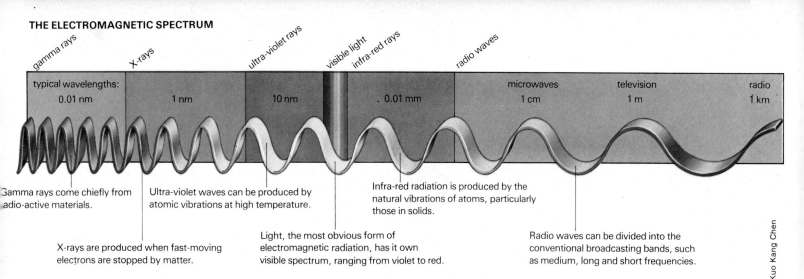

THE ELECTROMAGNETIC SPECTRUM

gamma rays	X-rays	ultra-violet rays	visible light	infra-red rays	radio waves		
typical wavelengths: 0.01 nm	1 nm	10 nm		0.01 mm	microwaves 1 cm	television 1 m	radio 1 km

Gamma rays come chiefly from radio-active materials.

X-rays are produced when fast-moving electrons are stopped by matter.

Ultra-violet waves can be produced by atomic vibrations at high temperature.

Light, the most obvious form of electromagnetic radiation, has it own visible spectrum, ranging from violet to red.

Infra-red radiation is produced by the natural vibrations of atoms, particularly those in solids.

Radio waves can be divided into the conventional broadcasting bands, such as medium, long and short frequencies.

revealed it in astonishing detail. Rocket flights carrying 'pinhole cameras' took the first pictures of the Sun in X-rays, and between 1962 and 1975 the US launched a series of eight Orbiting Solar Observatories, carrying X-ray detectors to study the hot corona. The real breakthrough, however, came with the American Skylab space station, launched in 1973. Its special solar observatory enabled the finest-ever pictures of the Sun's corona in X-rays to be obtained, and the US has kept up its solar research with the Solar Maximum Mission Satellite, launched in 1980.

It turns out that, unlike the Earth's atmosphere, the Sun's corona is not a uniform layer of gas, but is extremely patchy. In places the gases are concentrated into dense clouds, clearly under the influence of magnetic fields rising from the Sun's surface. Elsewhere, in the *corona holes,* the density of gas is low and X-ray emissions very weak. This is because the atmosphere is streaming outwards into interplanetary space as a 'solar wind' of electrically-charged particles.

The fact that the Sun is a powerful X-ray source is due only to its being so near the Earth. The normal corona produces less than one-millionth as much power in X-rays as the Sun emits in the form of light. Even during the outburst of a powerful flare, the Sun's X-rays amount to only one-thousandth of its light output, and this peak lasts only a few seconds.

Search for cosmic sources

Up to 1963, it was not generally thought that the universe harboured more powerful radiation sources than ordinary stars. But in 1962 Bruno Rossi and Riccardo Giacconi of the American Science and Engineering Corporation launched a rocket carrying three X-ray detectors, with the declared intention of looking for X-rays generated on the Moon by the impact of solar wind particles. In fact,

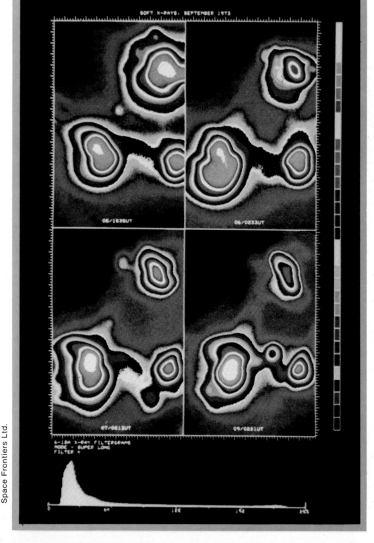

Left X-ray pictures of active regions of the Sun taken by the Skylab space station in 1973. Thermal imaging reveals the variation of radiation over the surface area. *Below* The soft X-ray telescope of the British satellite Ariel VI, which could be used to investigate the brighter X-ray sources not followed by the Einstein Observatory.

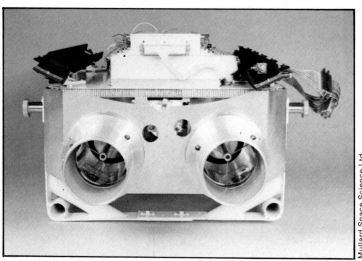

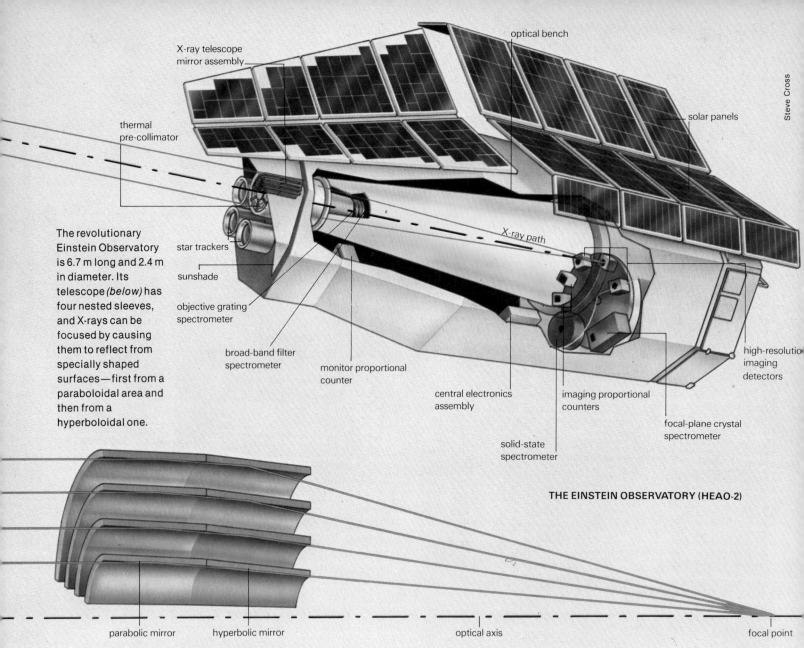

THE EINSTEIN OBSERVATORY (HEAO-2)

ASSEMBLY OF X-RAY TELESCOPE MIRRORS

they discovered bigger game: a distant cosmic X-ray source in the constellation Scorpius (a source now known as Sco X-1).

Within five years, several other sources were detected by rocket flights, with locations which coincided with objects visible to ordinary telescopes. Sco X-1 proved to be a distant blue star, emitting X-rays over a thousand times the power of the Sun at all wavelengths. A strong source in Taurus turned out to be the Crab Nebula, the remains of an exploded star or *supernova*; while an X-ray source in Virgo coincided with the massive galaxy M87, which lies in a cluster of galaxies 50 million light years away. Even in its infancy X-ray astronomy was revealing new kinds of violent astronomical objects, but rocket flights were too brief to study them in detail—an orbiting X-ray observatory was needed.

In 1970, the first X-ray astronomy satellite was launched off the coast of Kenya. Originally called SAS-A (the first of three Small Astronomy Satellites), it was rechristened Uhuru (Swahili for freedom). During its first day aloft, it clocked up more time observing the X-ray emissions than had been achieved by nearly a decade of short rocket flights.

Half a dozen more small X-ray astronomy satellites followed Uhuru during the 1970s. Astronomy satellites have only a limited lifetime: their batteries deteriorate; their electronics fail; or else they run out of the propellant gas supplies which controls the telescopes. Even so, most of the astronomy satellites have far outlasted their 'design lives'.

Some of the satellites are programmed to search the universe for new X-ray sources, others dissect the X-rays into a spectrum of wavelengths to reveal the temperature and composition of the gas emitting the radiation. Separate detectors are used to monitor the intensity—X-ray brightness—of individual sources. Most X-ray sources associated with stars in our galaxy, like Sco X-1, are not constant in brightness, but flicker continuously. The X-ray *pulsars* discovered by Uhuru emit regular pulses every few seconds. Another type of X-ray source—*bursters*—which discharge sudden irregular outbursts lasting about a minute were first detected by the Dutch ANS satellite. These transient X-ray sources appear un-

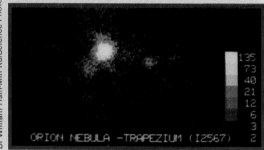

X-ray radiations discovered by the Einstein Observatory: Alpha Centauri *(left)*—Proxima Centauri is also seen in the inset; the star cluster of the Pleiades *(top)*; and the Orion Nebula and newly found Trapezium *(above)*.

predictably, and remain bright for several days before gradually fading over the succeeding months.

In the late 1970s, the US launched two huge X-ray satellites in the High Energy Astronomy Observatory (HEAO) series. HEAO-1 was essentially an enlarged, more sensitive version of previous satellites, but HEAO-2 was revolutionary. It carried the first real X-ray telescope, and enabled astronomers to 'see' for the first time the structure of the X-ray emitting sources.

Earlier satellites had used various types of *collimator* to restrict their 'view' to small regions of space at a time. Conventional telescopes had not been carried because X-rays are extremely difficult to focus. A lens cannot bend their paths to a focus without absorbing the rays, and if an ordinary dish-shaped mirror is used, as in an optical reflecting telescope, the X-rays are absorbed rather than reflected. X-rays are reflected only if they strike a polished metal surface at an angle of less than a degree.

The basic design for the X-ray telescope employs a 'sleeve' around the top end of the telescope tube. X-rays coming in close to the inner edge of the sleeve strike its tapering inner polished surface at a shallow angle, and are reflected to a focus a long way farther on. The amount of X-rays caught can be increased by stacking several sleeves one inside the other, each reflecting X-rays to the same focal point. The HEAO-2 satellite had four sleeves nestled together, the outermost 58 cm (23 in.) in diameter. The focal point, where electronic detectors could inspect the X-ray image, lay 3.4 m (11 ft) beyond.

After the successful launch of HEAO-2 in 1978, it was renamed the Einstein Observatory, in honour of Albert Einstein whose centenary occurred shortly after. This observatory revolutionized X-ray astronomy: the focusing power of its telescope not only showed details for the first time, but it could also pick out X-ray sources 500 times fainter than those previously detected. In terms of optical astronomy, it was as if Galileo had suddenly progressed from his first crude 'optick tube' telescope of 1609—less powerful than binoculars—to a modern telescope such as the one at Mount Palomar.

Numerous X-ray sources

The Einstein Observatory far outlasted its one-year design life, finally failing in April 1981 when its propellant gases ran out. Its sensitive detectors had found X-rays coming from virtually every kind of astronomical object—ordinary stars, galaxies and the very distant quasars. It spent relatively little time, however, looking at the brightest X-ray sources. These could be investigated by the less-sophisticated satellites still operating or about to be launched, and as Einstein closed down, the British Ariel VI and the Japanese Hakucho X-ray burst-detector satellites were still scanning the heavens.

The Japanese are planning a series of three small Astra satellites for launch in the 1980s, and possibly a large X-ray satellite for 1990. The European Space Agency is scheduled to launch its Exosat X-ray observatory late in 1982; although carrying two small X-ray telescopes, it is not as sensitive as Einstein except at the longest wavelengths. American astronomers plan an X-ray timing satellite for the late 1980s, to study the X-ray pulsars and bursters.

Before the Einstein Observatory had been launched, astronomers had confidently predicted that only a few types of stars would have very hot coronas like the Sun, expecting only a few ordinary stars to be detected as X-ray sources. But the Einstein Observatory found X-radiation from almost all the stars it investigated. The dwarf stars turned out to be the most surprising X-ray emitters. Proxima Centauri, for example —the star closest to the Sun—is a dim red star only one ten thousandth as bright as the Sun; but in X-rays, its outer atmosphere is a hundred times brighter than the Sun.

THE PRODUCTION OF X-RAYS IN SPACE

THERMAL BREMSSTRAHLUNG

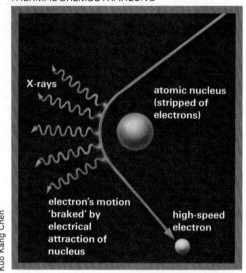

SYNCHROTRON PROCESS

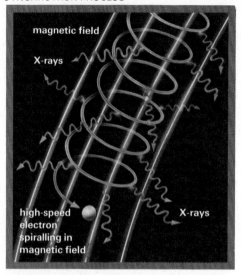

INVERSE COMPTON SCATTERING

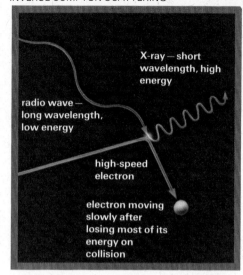

Left Thermal *bremsstrahlung* is the mechanism responsible for most X-ray sources; it occurs in very hot gases. *Centre* The synchrotron process occurs in strong magnetic regions, such as the Crab Nebula. *Above* Inverse Compton scattering may be an important source of X-rays close to black holes and in quasars.

These findings have necessitated a complete revision of theories explaining how coronae remain at such high temperatures. Previously, solar astronomers had thought that sound waves carried energy up to the corona, where their energy changed to heat; it now seems that magnetic fields convey the energy up to the Sun's corona in some more complicated way.

The Einstein Observatory also looked at stellar birth and death. Young massive stars in nebulae are prolific X-ray sources —probably because they generally have more active surfaces, with powerfully emitting coronae. After a massive star dies in a titanic supernova explosion it ejects a shell of gas at high speed as a *supernova remnant*. The shell expands outwards, sweeping up gas and heating it to a temperature of around ten million degrees. The hot gas shell appears as a beautiful luminous ring in the satellite 'photographs'. The Observatory's detectors could also pick out details of its X-ray spectrum which reveal the new elements made within the star and thrown out into space in the explosion.

Compact neutron star

The bright X-ray sources discovered by the early rocket flights and satellites originate from the collapsed core of supernovas. At this core is a *neutron star*, a collapsed ball of matter perhaps only 25 km (15 miles) across but weighing as much as the Sun. If in orbit around another star, the neutron star's powerful gravity can capture some of its companion's gas. This gas spirals inwards towards the compact star in swirling streams, forming a spinning *accretion disc* of gas. Friction within the gas heats it to around a hundred million degrees, at which temperature the copious X-rays are emitted by the inner parts of the disc.

This simple theory explains some sources such as Sco X-1, but others need further elaboration. The transient X-ray sources seem to consist of a neutron star in an elongated orbit, so that it draws gas from its companion only during the few days when the two are close to each other. X-rays then shine brightly, but as the stars draw apart, the gas flow stops, the accretion disc fades and gases spiral back to the companion star.

In other cases, the neutron star's magnetic field can affect the inflowing gas. In X-ray pulsars, the field directs the gas towards the neutron star's two magnetic poles, and the source bright in X-rays can be seen only when one of these 'hot spots' is facing us. As the neutron star spins, regular 'pulses' of X-rays are detected from the hot spots.

The magnetic field of the neutron star in a burster system would seem to retain a limited amount of spiralling gas suspended above the star's surface, until enough gas has built up to crush the magnetic field. As the gas drops suddenly to the surface, the shock of impact triggers a nuclear explosion which drives the gas out again with a powerful burst of X-rays.

There is one source, however, that fits none of these explanations. The brightest X-ray source in Cygnus, Cyg X-1, is undoubtedly a disc of hot gas surrounding a compact star in a double star system—optical astronomers have identified the ordinary star in the system, a bright blue star about 20 times heavier than the Sun. From the details of the orbit, the mass of the compact star can be calculated, and since it turns out to be as much as ten times heavier than the Sun, it cannot be a neutron star. The only object of

this weight yet small enough to have an X-ray-emitting accretion disc, is a *black hole*—the shrunken core of a star that has contracted to practically nothing. The gravitation field in the region that surrounds it is so strong that nothing—not even light—can escape. In Cyg X-1, the X-rays are produced by super-hot streamers of gas just before they disappear into the black hole.

The Einstein Observatory had the sensitivity not only to spot the X-ray sources in our galaxy—even millions of light years away—but also those in other nearby galaxies. Some of these sources are phenomenally powerful: one X-ray source, for example, in the neighbouring Large Magellanic Cloud galaxy, is a million times as powerful in X-rays as the total output of the Sun!

But the most interesting discoveries lie farther out in space, in sources a million times stronger again. These are associated not with individual stars but with whole galaxies or with clusters of galaxies. The Uhuru satellite had shown that distant clusters emit X-rays, and astronomers had concluded that huge pools of hot gas must lie between the galaxies in such clusters. The Einstein Observatory revealed in detail these giant gas pools, some over a million light years across and containing almost as much matter as the galaxies themselves.

The intergalactic gas has come from the individual galaxies, and has pooled together as they began to fraternize in the cluster. The Einstein Observatory found some clusters still aggregating from isolated galaxies, as shown by the fact that most of the X-ray emitting gas is still attached to the individual galaxies; at the other extreme, the galaxies in some clusters have lost all their gas, and it lies in a perfectly symmetrical pool through which the galaxies now swim.

Exploding galaxies

A few galaxies—both inside and outside clusters—have active 'exploding' centres. The British Ariel V satellite showed that many of these emit X-rays, and the Einstein Observatory has been able to pick up the most distant, and most powerful exploding galaxies of all—the *quasars*. It has found X-rays coming from every known quasar it has looked at—and astronomers have discovered many additional quasars by identifying previously unknown X-ray sources detected by the Einstein Observatory's telescope.

Quasars pack as much power as the output of ten million stars into a region no larger than our solar system. It is generally thought that they consist of a large accretion disc of hot gases spiralling into a massive black hole at the heart of a galaxy—a black hole as heavy as a thousand million Suns. The light and radio waves of a quasar emanate from the outer regions of the disc, but the X-rays must originate where the gas is hottest—just outside the boundary of the black hole.

The Einstein Observatory has helped X-ray astronomy develop from a frontier region of unexpected discoveries to a mature branch of astronomy, taking its place alongside optical and radio astronomy as a major contributor to our understanding the universe. It reveals the most energetic, most exotic places in the universe, from the Sun's still enigmatic corona to the last gasps of gas disappearing for ever into the massive black holes within the mysterious distant quasars.

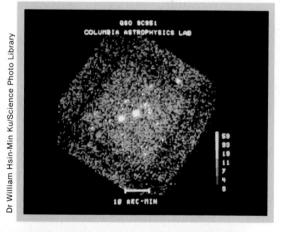
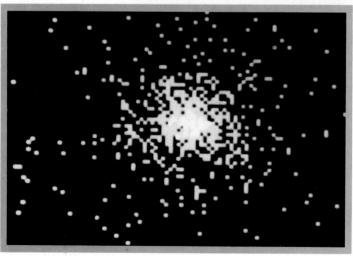

Left Artist's impression of the black hole in the Cygnus X-1 system. Matter is drawn off from the large star, forming a spiralling accretion disc around the companion star. The dots in the black hole image *(below)* represent X-ray concentration. *Right* Quasar found by the Einstein Observatory.

Forensic Science: Ballistics

Shot in the dark?

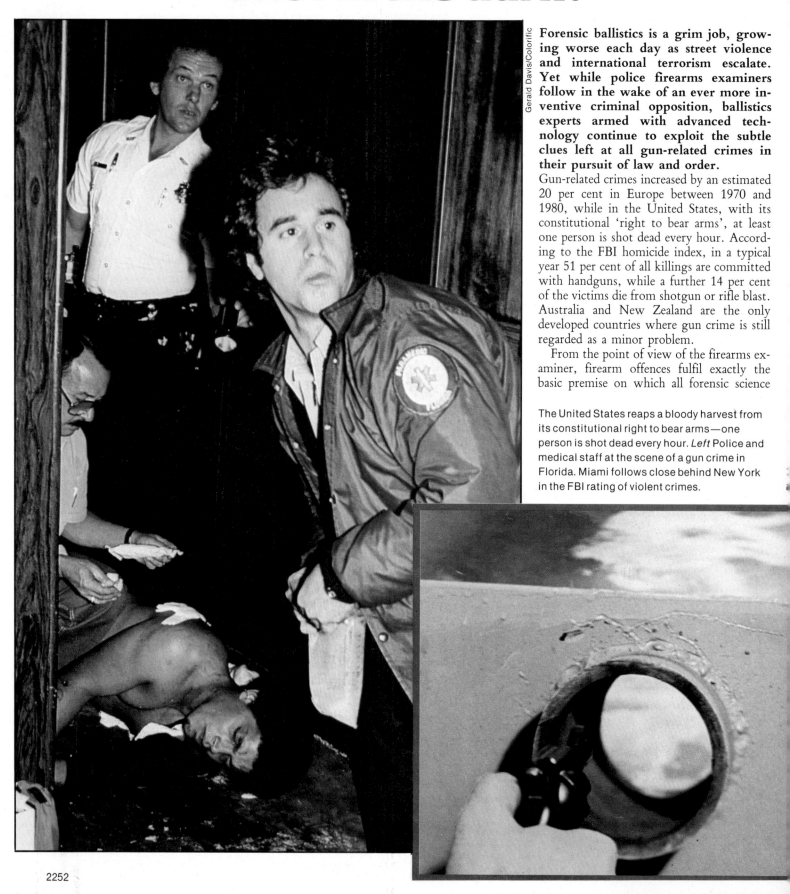

Forensic ballistics is a grim job, growing worse each day as street violence and international terrorism escalate. Yet while police firearms examiners follow in the wake of an ever more inventive criminal opposition, ballistics experts armed with advanced technology continue to exploit the subtle clues left at all gun-related crimes in their pursuit of law and order.

Gun-related crimes increased by an estimated 20 per cent in Europe between 1970 and 1980, while in the United States, with its constitutional 'right to bear arms', at least one person is shot dead every hour. According to the FBI homicide index, in a typical year 51 per cent of all killings are committed with handguns, while a further 14 per cent of the victims die from shotgun or rifle blast. Australia and New Zealand are the only developed countries where gun crime is still regarded as a minor problem.

From the point of view of the firearms examiner, firearm offences fulfil exactly the basic premise on which all forensic science

The United States reaps a bloody harvest from its constitutional right to bear arms—one person is shot dead every hour. *Left* Police and medical staff at the scene of a gun crime in Florida. Miami follows close behind New York in the FBI rating of violent crimes.

rests: that every contact leaves a trace. In fact, this is triply so in the case of a shooting. For there is contact between the killer and the gun, the gun and the bullet, and bullet and its victim. Forensic firearms examiners are concerned with the analysis, identification and comparison of all the traces left by these contacts.

Both the police and the press habitually use the term 'ballistics' to refer to firearms examination departments. Though useful, this is a misnomer as the word properly means the study of missiles in flight. Although a knowledge of bullets and their trajectories, calibres and loads is a major part of the job, every aspect of the three-fold contact in a shooting needs specialist examination. For instance, because the wound is so intimately connected with the bullet, the firearms examiner will also become something of an expert on forensic pathology.

The investigation of major crimes such as homicide may eventually involve hundreds of police officers from all branches of the service, but initially, six people are central to the case. These are the detective in overall charge, known as the investigating officer, the photographer, the finger-print officer, the pathologist, the scene of crime officer—whose duty covers the immediate search for material clues and the collating of evidence—and the forensic scientist whose speciality is appropriate to the crime—in the case of a shooting, the firearms examiner. Each may contribute a piece to the picture but the firearms examiner finally fits together the scientific jigsaw.

In the case of a homicide the first concern will be with the body. Obviously if a weapon is present, there is always the possibility of a suicide or accident. The firearms examiner and the pathologist must first rule out these alternatives, before beginning a full scale murder investigation. Certain near-standard indications will help them. A suicide usually—though by no means invariably—goes for a head shot, placing the pistol either to the temple or into the mouth. Shotgun suicides surprisingly often remove a shoe and sock and press the trigger with their toes. In many cases, a contact head shot, with a low calibre pistol, will literally blow out the brains. These are extruded from both the entry and exit wounds by the sudden build-up of pressure in the vault of the skull, caused by the expanding gases and the 500th of a second passage of the bullet. There may also be a bruise at the edge of the entry wound caused by the recoil of the weapon, along with scorching and 'tattooing' of the skin from powder and other residues being embedded there.

But accidents can also result in contact

Stages of weapon analysis at the FBI lab in Washington DC. *Below far left* Water tank used to obtain a bullet for comparison with one used in a crime. *Below left* A ballistics expert examines an auto pistol for distinguishing marks. *Below right* The FBI's weapons room.

wounds—falling on to a shotgun for instance—as of course can murder. In the case of apparent accident, the state of the weapon's mechanism may tell its own tale. Worn or faulty locks and safety catches frequently cause devastating injuries when a shotgun is carelessly handled. By and large, however, the investigation of a suicide or accidental shooting is straightforward enough. The position of the body in relation to the weapon provides a strong indication of what has occurred, and the position and identification of finger-prints on the gun may clinch the matter beyond a doubt.

Murder clues

A case of murder is more complex, often vastly so. The first stage in its investigation will be an attempt to reconstruct the crime as far as possible. The size and appearance of bullet wounds often gives a fair indication of the distance and direction from which the shots were fired. Where a shotgun is concerned, the firearms examiner can reproduce the pattern and spread of the pellets found on the body on a firing range by shooting at blank white cards. When the pellet holes on the cards match those on the body, the distance has been established.

The gauge of a shotgun can be estimated to some extent by the size of the wound, but discovery of the wad which encloses the end of a shotgun cartridge will give the gauge exactly. Sometimes this is driven into the wound and is recovered during the autopsy.

INSTRUMENTS OF DEATH

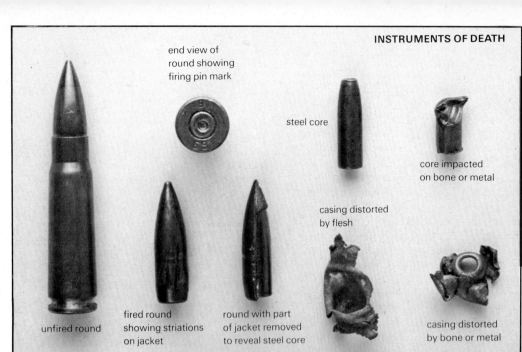

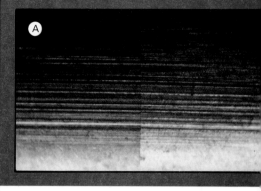

Left A rifle round from a Soviet Kalashnikov AK-47 before and after firing and on meeting different types of target. It has a steel core, soft lead tip and cupro-nickel casing.
Below Same type of round in guerilla warfare victims. *Left* Hand wound. *Right* Chest wounds.

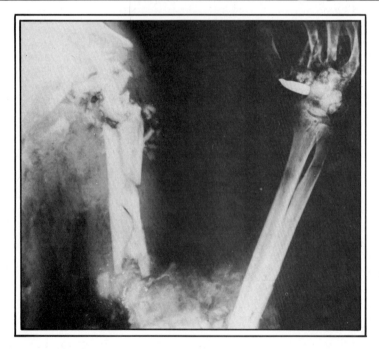

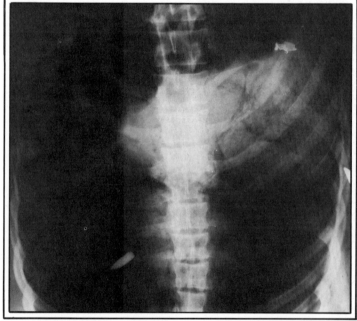

But unless the murderer has been particularly cool and methodical, it will be lying around somewhere. After estimating the direction and angle of the shot the firearms examiner will assist the scene of crime officer and the crime squad in their meticulous search for the wad, stray pellets or bullets, and discharged cartridge cases. Where a killing has occurred in the open, this task can be a long and tedious one, but indoors the missiles may be found embedded in walls or furniture. Where possible they are cut out complete with the material in which they have lodged so as not to damage them. As each item is found, the photographer takes pictures showing its position in relation to the overall scene, and the scene of crime officer marks it on a graph paper map. This process is carried on before anything is moved or touched at the murder scene.

Next comes the examination of the firearm itself. Normally a murder exhibit is left as far as possible in the condition in which it is found, but a firearm is a notable exception. Firstly it is examined, untouched, with a hand lens, and in the case of a revolver, the position of the bullets in the chamber is noted. By shining an indirect light on the gun, the finger-print officer may spot latent prints, and these are delicately dusted and photographed as the weapon lies. Then the gun is carefully unloaded, the magazine, if it has one, removed, and the cartridges extracted and sealed in individual plastic bags. A separate bag is placed over the muzzle of the gun to preserve any powder residues that may remain.

The recommended methods of picking up a handgun are to wear a plastic glove and grip the gun firmly by the checkered part of the stock or butt, which is unlikely to hold finger-prints, or to hook a piece of stiff wire through the lanyard ring where fitted. Not, as seen in countless films, by inserting a pen down the barrel or through the trigger guard: practices both damaging to evidence and hazardous to life!

Another 'television cop' habit—sniffing

Forensic Science: Ballistics

Left The identification of either a bullet or a gun connected with a violent crime is one of the most basic functions of forensic ballistics experts and is termed the *bench* stage of the investigation. Comparison microscopes have been used since the 1920s for this task and microphotographs record the results. A is a microphotograph of two bullets, B shows two sets of breech face marks and C compares two firing pins. All are from the FBI Academy's collection.

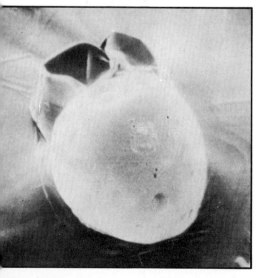

Views down a scanning electron microscope. *Above* The inside of a gun barrel showing rifling 'lands'. *Below* Firearms particles on tapings from a hand—magnified 6,500 times.

the gun barrel—provokes amusement rather than horror in real life murder investigators, though it is dangerous as well. Sniffing the gun barrel will tell the examiner only that the gun is not brand new, for the scent of cartridge discharge can linger indefinitely in any gun not thoroughly cleaned.

At this stage the scene shifts to the ballistics laboratory itself, where the examiner will begin bench work on the clues to hand. Besides its test firing range—usually 15.5 m (25 yards) in length and nowadays equipped with static air sampling pumps to protect examiners from dangerous lead fumes—the laboratory will contain horizontal bins filled with water into which test shots can be fired without the bullet being distorted by impact.

Much practical ballistics involves microscopy and there are three instruments essential to this branch of forensic science: the comparison microscope, the fibre optic bore microscope used for viewing rifling details and the measuring microscope for calculating rifling characteristics on fired bullets.

The comparison microscope basically consists of two lenses close to the subject joined by a simple viewing lens. This enables inspection of two bullets for correspondences. For forensic work both of these instruments are equipped with camera attachments for recording results.

The firearms laboratory also contains chemical benches for testing residues found in clothing, guns and sections of human tissue, and a metalwork area for gunsmithing. The comparison of bullets and guns plays a major part in ballistics, and firearms examiners have access to ballistics control centres such as the Home Office Central Research Establishment at Aldermaston, England, and the much larger FBI Forensic Ballistics establishment in Washington DC. These have every make of firearm and cartridge on computer file along with specimen weapons and bullets, and they handle enquiries from many parts of the world.

However, experienced scientists also have personal 'data banks'—knowledge which indicates immediately what to look for, and in which direction. They might begin with the fact that all guns divide into two types—rifled and unrifled. Shotguns and some airweapons are smooth bore while most other guns have spiral grooves cut into the interior of the barrel to spin the bullet and keep it on course. Manufacturers vary in the number and directions of their rifling grooves and corresponding *lands*—the raised sections between the grooves.

When a bullet passes down a rifled gun barrel, the rifling leaves *striations*—marks on the metal—and if a relatively undamaged bullet is recovered, the examiner will be able to make a fair guess at its manufacturer—from analysis of the metal itself—the calibre of the gun from which it came and, by checking the direction and number of striations, the model and make of gun.

Stray cartridges

Along with ordinary shotguns, the firearms most commonly used by criminals are handguns—revolvers, automatics or sawn-off shotguns.

When any firearm is discharged, a hammer or firing pin strikes a percussion cap at the base of the cartridge, igniting the main explosive which powers the bullet. A tremendous force is released when the cartridge 'goes off'—nearly five tonnes per square centimetre (12 tonnes per sq in.) in some handguns—which expands the cartridge case against the walls of the firing chamber and simultaneously slams it back against the breech face. By stripping the weapon down and examining the firing chamber and breech face under a microscope, any tooling marks on these surfaces can be checked for comparison with ejected cartridges, for the force of firing will have imprinted any such marks on to their surface.

Forensic Science: Ballistics

For practical forensic purposes, a shotgun or revolver can be difficult to pin down, for unless the killer has reloaded, the discharged cartridge will remain in the weapon and be carried from the scene. With an automatic or pump action shotgun with the pump in operation, cartridges are ejected, usually being thrown high and to the right. This means that even the most assiduous murderer has to search for them to gather them up, and with surprising frequency they are thrown on to the tops of high furniture, and overlooked.

Because a self-loading pistol, shotgun or rifle has more moving parts in contact with the shell, an ejected cartridge from one of these weapons will bear a number of marks, each unique, from the firing pin, ejector, magazine, slide, bolt face, chamber wall and extractor. Collectively these marks will identify the gun which made them.

Super glue

The final part of the three way contact involved in a shooting is that between the murderer and the gun. If the weapon has not been thoroughly wiped there may be latent prints on both the gun and the bases of any unfired cartridges, though for various reasons these are difficult to 'lift' in a clear enough manner for presentation as anything but circumstantial evidence. However, a recent development, the result of an experimental 'spin-off' by two scientific police officers from Northampton, England, has been the use of vapour from *cyanoacrylate*—super glue—as a finger-mark reagent on a variety of difficult surfaces. Using this method, clear and usable prints have been obtained from plastic shotgun cartridges—a significant boost to forensic ballistics.

Experiments are also underway to examine the properties of *blow back*—residues, often barely visible, which issue from the muzzle and breech area of a gun when it is fired. Until recently, dermal nitrate tests were carried out on the hands of a suspect. Unfortunately, the test can only eliminate suspects for nitrates are also present in cigarettes, cigars and various everyday chemicals such as fertilizers, but the test is nevertheless still in use in some countries. Swabs are still taken from the hands and clothing of the suspect, however, and compared with those taken from the victim's wounds and clothing. Sodium rhodizinate is used for testing the presence of lead or barium, but usually only for elimination.

An almost prohibitively expensive test, neutron activation analysis, has been tried from time to time in the United States. A thin 'glove' of paraffin wax is set on the suspect's hand and then peeled off and placed in the heart of a nuclear reactor until it is radioactive and gives off gamma rays. Using an instrument called a gamma ray spectrometer, the scientist sorts and measures the rays to show what elements are present. If mercury, lead, antimony or barium are present, it is highly likely that the suspect has recently fired a gun. In fact, this test proved inconclusive when tried on Lee Harvey Oswald, President Kennedy's assassin.

However, the phenomenon of blow back is potentially a valuable source of forensic information. In 1978 the Home Office contracted the British Chemical Defence Establishment at Porton Down to use their expertise in high speed photography to examine the trajectory of bullets and the distribution of gases at the moment of firing. Their pictures, taken at 1/3,000 second intervals, have been used by the Home Office's own Research Establishment at Aldermaston in a series of secret experiments to make blow back evidence a positive weapon in the forensic science armoury.

Home-made guns

Since its inception in the 1910s, forensic ballistics has tended to depend on observation rather than theory and seems likely to remain this way. For instance, the growing use of reproduction weapons altered by amateurs to fire bullets has meant that firearms examiners have had to start from scratch in a new field, building up data on the characteristics of such home-made guns —including conducting experiments of their own. Nevertheless, the dramatic rise of gun crime means that ballistics, the youngest of forensic sciences, is rapidly coming of age.

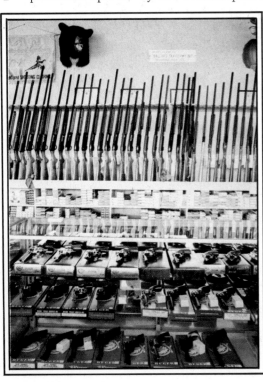

The American public still has ready access to guns at shops like Dawson's in Annandale, Virginia *(left)* despite protests triggered by the wave of political assassinations after President John Kennedy's death in 1963. *Below* J.F.K. slumps foward after the first shot.

Forensic Science: Art Forgery

Exposing the forger's art

With old master paintings commanding vast sums of money in auction rooms throughout the world, establishing a picture's authenticity has become vitally important—and it is to the scientist that this onerous but intriguing task has fallen. Combining art appreciation with laboratory techniques, the work of forensic scientists has vastly increased our understanding of our most highly prized works of art—and it has turned up a few surprises too!

The attribution of paintings to artists often long dead demands great skill and judgement, for a painting which first saw the light of day on the artist's easel several hundred years ago may have been subjected to numerous copies and imitations from the very first days of its existence. It may have been cropped to a smaller size or had new sections added by one of its subsequent owners. Areas of its surface may have been repeatedly replaced or overpainted to cover damage or to bring an old work up to date.

In a world where a million dollars may rest not on a painting's quality but on the artist who painted it, the need for a scientific approach to painting identification is clear.

Scientific analysis has proved that *The Fortune Teller (above)* is no amateurish recent forgery, and many consider the findings proof-positive of the picture's authenticity.

Two such cases, which came to the attention of art critics and collectors worldwide, concerned the artists John Constable and Georges de la Tour. These illustrated both the capabilities and limitations of forensic science when applied to art.

The discovery that certain paintings popularly believed to have been painted by John Constable were, in fact, painted by his

Forensic Science: Art Forgery

Near Stoke-by-Nayland (left) was long regarded as a fine example of John Constable's work. Reattribution of the painting to his son Lionel is reflected in its saleroom value.

Infra-red and ultra violet light have long been used in forensic science for the examination of paintings. Greater sensitivity can now be achieved with reflectography, which is used by many major art galleries. A vidicon television camera *(right)* is fitted with a specially designed infra-red pick-up tube.

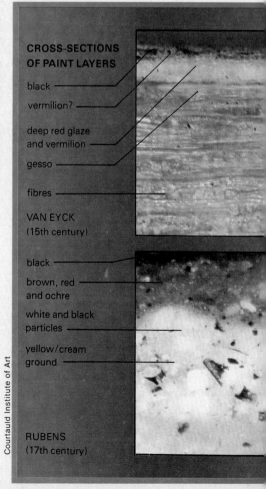

CROSS-SECTIONS OF PAINT LAYERS

- black
- vermilion?
- deep red glaze and vermilion
- gesso
- fibres

VAN EYCK (15th century)

- black
- brown, red and ochre
- white and black particles
- yellow/cream ground

RUBENS (17th century)

son Lionel was the product of meticulous detective work combined with subjective analysis. As the two artists had broadly similar styles, were contemporaries and shared materials, scientific comparisons of their work could, at best, only have produced a tentative conclusion that certain paintings were the product of two artists.

Scientific analysis did, however, play an important role in establishing certain facts about Georges de la Tour's *Fortune Teller*. The painting was bought by the New York Metropolitan Museum of Art in 1960, less than ten years after it had been 'discovered' in a French château. Despite its mysterious past, the painting was considered unquestionably genuine and was put on display to the public. Then, in 1980, art critics Christopher Wright and Diana de Marly drew attention to certain features which they believed indicated the painting to be a fake. They claimed that the painting was well below de la Tour's standard and that the costumes of the figures portrayed were inconsistent with both painter and period. The appearance of the French obscene word *merde* and a dubious signature further fuelled the fires of doubt.

Somewhat embarassed, the Metropolitan Museum of Art turned to its science department. Superficial analysis clearly showed that the word *merde* was a later addition—perhaps a bizarre joke by a restorer—and that the painter's signature had been obliterated, possibly during a period when the artist was out of favour.

Further hard evidence came from the analysis of pigments used in the painting. Pigments such as lead white which are consistent with a 17th century origin were identified as was lead-tin yellow, a pigment not used since de la Tour's day. Supporting the laboratory evidence, documentary evidence was discovered which proved that the painting was in existence in the 19th century. The painting is now back on show and the museum curators feel vindicated—though the critics are not convinced.

Of the techniques used by forensic scientists, pigment analysis is one of the most potent. On many old paintings the paint is hidden by a thick, discoloured layer of varnish. But in most cases, the exact composition of any part of a picture and each of the various layers of pigment can be identified by its different chemical makeup.

The removal of paint for analysis—even though samples are typically less than 1 mm in diameter—is done only on larger paintings. But by analyzing a wide range of work, gallery data banks are being established with records of the types of pigment used by various artists throughout their careers. As this data becomes more complete the task of identification will be significantly simplified.

Paint samples are taken with either a scalpel point or a cut-off hypodermic needle. The sample is then placed in a mould with a transparent material such as polyester resin, which sets hard. The plastic is carefully ground away until a cross-section is revealed—consisting of the various layers of paint, ground and varnish—ready for microscopic examination. Under the microscope, it is possible to see how many layers of paint there are and to detect the layers of different colour. Often, however, techniques of chemical analysis are required, particularly when identifying pigment base or *medium* which is likely to be an organic substance.

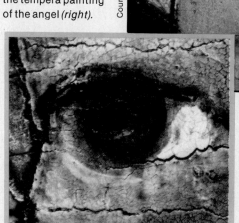

By photographing a small part of a picture at intense magnification, small changes in the structure of the brushwork may be revealed. The enlargement of the eye *(below)* shows the green underpaint in the tempera painting of the angel *(right)*.

For further analysis gas chromatography is a valuable technique. Tiny samples of paint are vaporized and the gas produced is passed through a tube of inert material. This material impedes the different constituents of the gas at different rates. As the separated substances emerge from the tube they are recorded as peaks on a graph, producing a blueprint of the whole sample. The height of the peaks on the graph indicates the probable identification of each substance and together make up a pattern unique to that type of material. In this way, many common types of paint can be spotted relatively quickly.

Another useful technique uses a laser beam to vaporize minute areas of the paint cross-section—effectively isolating each layer for individual analysis. The cross-section is placed in a special microscope and the laser is aimed at a target layer. The tiny portion of gas which is produced passes between two carbon electrodes to discharge; the light that is then released has its spectrum characteristics recorded on a photographic plate. In this way, the chemical elements present in a single layer can be identified.

The examination of paint under a microscope *(left)* may reveal an artist's underpainting or a change of mind, and will even show subsequent, possibly fraudulent, overpainting.

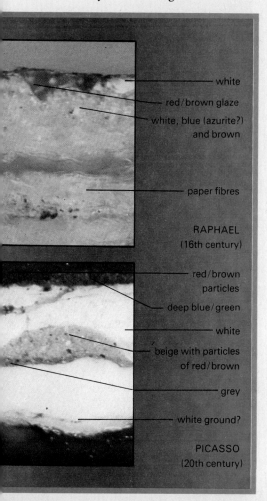

RAPHAEL (16th century)

- red/brown particles
- deep blue/green
- white
- beige with particles of red/brown
- grey
- white ground?

PICASSO (20th century)

- white
- red/brown glaze
- white, blue (azurite?) and brown
- paper fibres

As the array of sophisticated laboratory equipment grows, it is easy to overlook the value of a thorough visual inspection when examining a painting. Nevertheless, basic facts such as the support used (canvas, wood, copper, plaster), the type of paint (oil or water-based) and whether the painting has been cropped or damaged can soon be established by an expert with a meticulous eye for all these details.

The next technique used employs *raking light*. By viewing the picture in a darkened room with light from a single source falling across the picture surface at a shallow angle, minute variations that would be invisible in normal light become apparent. Subtle differences in the relief of canvas, for example, may show joins between different pieces of material, or repairs to tears.

Detecting damage

Sometimes, if an old picture has at sometime been transferred from wood to canvas, the grain of the original wooden support will show. Raking light is most useful in revealing subsequent overpainting which might hide the technique of a master. Flaking paint and missing fragments also show up, as does loose or stretching canvas.

Both raking light and normal light can be used in conjunction with photographic enlargement. The maze of minute cracks known as *craquelure*, which cover the surface of most old pictures, will show up clearly in the magnified image. Abrupt changes in the patterns made by the cracks indicate a change in the type or age of the paint medium, again revealing later alterations and providing evidence about the picture's authenticity. Inscriptions are particularly likely to have been tampered with—as in the case of *The Fortune*

Forensic Science: Art Forgery

Teller—either to disguise the signature of an imitator of an important artist or even to add a false attribution. Close photographic inspection can show whether the signature is original—showing whether the craquelure continues through the signature or the signature has been painted on top.

Analysis using other, invisible, forms of light can further aid the forensic scientist. The ultraviolet illumination of a picture in a darkened room, for example, produces a quite surprising result. Using the light from a mercury vapour lamp, filtered by a special nickel oxide filter so that only the invisible ultraviolet wavelengths are passed, causes the surface of a picture to fluoresce. In this case different materials, not different colours, respond with varied brightness.

Old varnish takes on a milky translucence. Paint added above the varnish appears dark and can indicate retouching or damage repair. Without the varnish, however, the technique has limited usefulness, for while the different areas of paint fluoresce to different extents, the result is very difficult to interpret. An exception is the pigment zinc white, which is easily identified and, as it was only introduced in the late 18th century, its presence in a picture purporting to be older would give rise to the suspicion of repainting or forgery.

Ultraviolet fluorescence's widespread use in examining pictures has led to the applications of cellulose acetate varnishes—which have a totally opaque appearance under ultraviolet light—by those wishing to disguise serious damage to a picture. Its greenish appearance can be recognized, however, indicating the need for caution.

Infra-red light is also used. Although infra-red radiation is invisible to the naked eye, special photographic films record it to the exclusion of visible parts of the spectrum. Because some substances are more or less transparent to infra-red light, it is able to penetrate varnishes and other surface layers of paint—sometimes revealing hidden layers below, and sometimes even the artist's original underdrawing. Pictures which are very difficult to view because of thick, discoloured varnish can be photographed on infra-red film to reveal hidden detail.

Infra-red reflectography

Because infra-red photography penetrates the varnish layers, unlike ultraviolet light, it can be used to detect overpainting and old restoration work beneath subsequent surface applications. This makes it useful to restorers identifying and removing parts of a picture which are not original.

Infra-red photography is now augmented by infra-red reflectography, introduced by J. R. J. Van Asperen de Boer in Amsterdam and now used by a number of major art galleries. An infra-red vidicon television camera is used, with the infra-red image displayed on a closed-circuit television. The technique is faster than photography, and is more sensitive to longer wavelengths in the infra-red part of the spectrum.

The most dramatic results of all are, however, produced by X-radiography. X-rays have been used to penetrate the mysteries of old pictures for most of the 20th century. Most major museums have X-ray facilities and are gradually working through their collections building up comprehensive libraries of X-ray information. X-ray plates are placed as close as possible to the picture surface and exposed to X-radiation at low power from the rear of the painting. With the exception of copper panels which are impervious to X-rays, the result is a complex image of all the layers of paint and the material on which the picture is painted.

A famous example of X-radiography leading to the reappraisal of a major work concerned the portrait of Pope Julius II in the National Gallery, London. This celebrated work is known in several versions, and the original by Raphael was believed to be in the Uffizi Gallery in Florence. The London picture was generally considered a very old copy, perhaps even produced by the pupils of

The painting of a nude by Modigliani *(far left)* can be compared with the same painting viewed under raking light *(centre left),* which shows up minute canvas variations, and by X-rays *(left),* now widely used to confirm disputed attributions. Such a case was the portrait of Pope Julius in the National Gallery, London *(below left* and *below).*

the artist under his supervision. In 1969, it was subjected to X-ray examination, with startling results.

One of the most obvious characteristics of a copy is that the artist or forger has no significant changes of mind while he is painting it—he has a complete image in front of him and he merely reproduces it as closely as he can. The National Gallery Julius II showed no such certainty, however. The X-ray clearly revealed an artist experimenting and changing his mind in the course of painting the picture. The backcloth, a plain green in the finished picture, is seen on the X-radiograph to have been decorated with a pattern of papal tiaras and also a different pattern of crossed papal keys, both ideas being abandoned in favour of the plain green curtain. On the other hand, the acorn knobs of the chair seem to have been painted over a simpler version beneath.

National Gallery original

Other, even more significant, changes can be seen on closer inspection. One chair knob has been moved further to the left from its original position, the handkerchief the Pope holds was once much larger or perhaps even a document, and even the Pope's facial features were experimented with. It is now certain that this version is the original from which the others were copied.

In some cases, pictures not even known to exist have been discovered by X-rays. Rembrandt's portrait of his son Titus, for example, covers an earlier work—a virgin—by the artist. Valuable knowledge about the evolution of a work is also often obtained. Manet's *Olympia,* for example, has much coarser features in the underpainted version seen in X-ray, showing even more shockingly than the finished work that the artist's goddess is supposed to be recognized as a Paris prostitute.

Technical details can also be shown, adding to the understanding of a picture's history and origin. Old paintings on canvas will almost always have been lined with new canvas. X-rays can reveal the original, yielding valuable information from the size and pattern of the weave. Sometimes, paintings have been transferred bodily from one canvas to another or from wood to canvas. Again, X-rays can produce an image of the original canvas on the back of the old paint.

The institutions most active in the scientific study of pictures are rapidly extending the range and sophistication of their analytic techniques, including the most advanced forms of microchemical analysis. As data is gathered from a greater number of pictures, so more is becoming known about the techniques used by painters in different ages.

Thanks to the scientists, this mass of objective evidence about a picture's total composition now makes it possible to attribute a work to a particular artist and period with greater confidence. Furthermore, it makes the detection of older forgeries and modifications to paintings easier, and sets an almost impossible task to the modern forger.

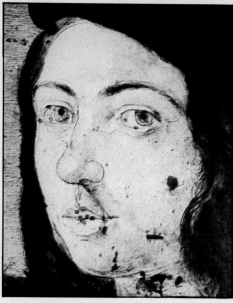

Normal colour reproduction of *Portrait of the Artist* by Raphael *(top),* and the same painting seen under ultraviolet light *(centre)* and infrared light *(above).*

Resources: Food

Soya: food of the future?

Soya is one of the great success stories of the 20th century, yet even today its potential is still not fully exploited. It is such a cheap and excellent source of protein that if food production and eating habits could be changed to incorporate it fully into the world's diet, there would be a protein surplus in the world instead of the current pattern of shortages and gluts.

For food technologists soya is one of the most exciting raw materials available today: it can be refined to almost pure protein with a flexible structure that can be built up to resemble not only meat but entirely new concepts in food.

The soya bean has been cultivated and eaten in South-East Asia for millennia but modern developments in its use have had most impact on the agriculture and economy of the USA. The soya plant was first introduced to the USA in the early 19th century and grew well in the southern and midwestern states. But for 100 years most research was based around its use in increasing soil fertility and as a silage constituent.

The breakthrough discovery of its oil and protein qualities occurred in the first decade of the 20th century and the first commercial pressings of oil and meal-cake were made in 1911. Now it is the single most valuable commodity in the USA where 70 million acres given over to soya supply 60 per cent of the world requirement. Brazil is the second major supplier and China the third.

The soya plant was brought over to Europe in the 17th century by travellers but failed to flourish because of the cold and frequent early summer droughts. It needs a damp, warm climate and plentiful water supply which is drawn up by a long, branching taproot penetrating deep into the ground. It usually grows about 1 m (3-4 ft) high, though irrigation can produce plants of over 1.5 m. The hairy pods contain up to four pea-sized round or oval beans which can be green, brown, black, mottled or the yellow types favoured for oil.

Many people eat soya regularly without knowing it. *Right* These meat products contain soya refined into either spun or extruded Textured Vegetable Protein (TVP).

All pictures in this article are by courtesy of the American Soybean Association

Soya beans are a dual raw material—after the oil has been removed, the residue is a high-protein meal which can be processed for humans or animals. Soya bean oil now accounts for one fifth of the world's total edible oil supply but the most exciting aspect of soya development is in direct human food.

Meat analogues

Food ranges from traditional products such as the bean curd known as *tofu* and fermented pastes to the newer protein-rich flours and meat analogues. These imitation meats—ersatz bacon rashers, 'beef' chunks or mince extender—have a very special role to play in defusing inflation in developing countries. Meat consumption is one of the first things to rise with higher incomes and as meat prices escalate, other commodities follow—giving rise to the maxim 'Inflation feeds on red meat'. In Saudi Arabia, for instance, meat consumption is currently rising by an annual 8 per cent. Cheap, acceptable meat substitutes could break this cycle.

Protein from soya is far cheaper than from animal sources. A soya-based bacon analogue requires one tenth the land used by pigs to produce an equivalent amount of real bacon and a beef analogue twenty times less than real beef. Soya plants have an exceptionally high cropping ability—a hectare produces 2,000 kg of beans which yield 400 kg of oil and 740 kg of crude protein. Pollution is also reduced because the soya plant fixes atmospheric nitrogen in the soil, whereas grazing land requires large amounts of fertilizer.

Moreover, from the food scientists' point of view, soya is by no means the poor relation of meat. Their enthusiasm for its nutritional value lies in its high protein content—the soya bean contains some 40 per cent protein by weight, more than twice the level found in meat and fish. This protein contains the eight amino acids essential for the human diet in highly usable proportions, although the exact amount varies according to the stage of processing.

Traditionally, soya beans are processed in two ways in their countries of origin. Soybean milk is produced by treating the ground beans with hot water. Further treatment with calcium salts precipitates the protein and oil together in the form of a bean curd. Cooked soya beans can also be fermented to yield a number of products including soy sauce and miso paste.

The modern industry uses an extensive processing cycle which delivers various products at various stages. Culled and separated by huge harvester-threshers, then stored in airtight containers, the beans are first screened for size and air-cleaned on arrival at the processing plant. Then they are passed through corrugated rollers which crack them

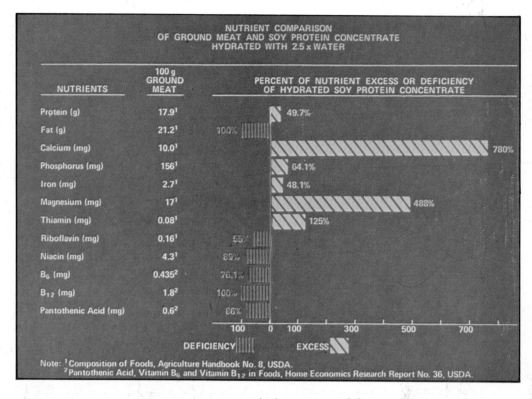

Above This nutrition table explains why soya is an excellent substitute for meat.
Below Scientists constantly research new strains of soya for different climates.

Resources: Food

into pieces, simultaneously separating off the dry outside hulls. These hulls are diverted to be heat-treated and milled to make soya bran. The bran can be incorporated into cereal foods and crisp-breads, though most of it goes into animal feed compounds.

Some of the cracked beans are toasted, cooled, and then ground up into full fat soya flour. This has a protein content of about 40 per cent; it is used in the baking industry, and also as a protein supplement in aid programmes. In baking, a small amount added to bread bleaches out the yellow pigment in wheat, improves the texture, and helps the bread stay fresh. As a diet supplement, full fat soya flour can be mixed with wheat, corn and milk solids, taken as a hot drink or soup, or baked into biscuits and buns.

Oil extraction

Most of the cracked beans have the oil extracted from them. First they are made into flakes by being steamed in giant kettles to soften them and then squeezed into very thin flakes between large-diameter rollers.

The flakes are then transferred to an extraction tower where they are moved around on wire trays and sprayed with a solvent, usually hexane, which dissolves the oil out of the solids. The mixture of hexane and oil washes down through the wire trays to a collecting tank where it is heated to the point where the hexane gasifies.

Crude soya oil remains in the tank and to refine this, gummy residues are removed by

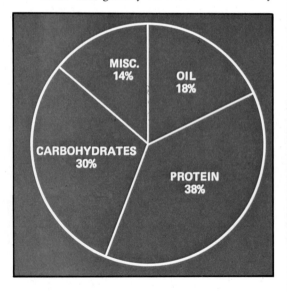

Above Soya beans have much to offer the human body. Apart from protein for tissue regeneration, plus carbohydrate and oil for energy, they contain vitamins A, B and C plus vital minerals and trace elements.

Left Soya is planted in rows on highly mechanized farms. *Top left* Shoots are strangled by weeds unless well hoed. *Above* Soya flowers contain female and male organs for self-fertilization. *Below left* The soya pod bursts when beans are ripe. *Top right* Harvester thresher unloads beans for transport to elevators (below).

mixing the oil with water or a weak salt solution. The oil rises to the surface, leaving the gums and other impurities in the watery layer. The gums are dried, producing lecithin which is used as an emulsifier in rubber, pottery, textiles, cosmetics and foods.

The next stage is the removal from the oil of fatty acids which could render it liable to go 'off'. It is reheated in tanks, sprinkled with an alkali solution, and stirred with paddles. The fatty acids combine with the alkali to form a thick soap solution which is easily separated from the oil and utilized in the soap industry. Any soap traces are removed from the oil by washing and vacuum drying. While the oil is still under vacuum it is finally bleached by the addition of a special type of Fuller's earth.

The clear, refined oil is pumped away for *hydrogenization;* this entails treating the oil with hydrogen gas at high temperature under pressure in the presence of a nickel catalyst. Light hydrogenization makes the oil more stable, though still liquid. Further hydrogenization makes the oil turn solid at room temperature, and is used in preparing oils for margarine production. The hydrogenized oil is now *winterized*—the removal of fats which crystallize at ground freezing point. This gives a salad-grade oil that will not cloud when chilled. It is then deodorized by blowing steam through the hot oil at high pressure.

When the oil has been extracted, the protein-rich flakes which remain form the basis of the most important of all high protein animal feeds. And 85 per cent of flakes are made into cake and meal for livestock.

The introduction of soya into animal feeds has brought great increases in final yields. Cattle which added 3.6 kg (8 lb) in weight for every 45.4 kg of feed 40 years ago now add 7.2 kg for the same intake and milk production has virtually doubled

To make soya foods for human consumption, the defatted flakes not going into animal feeds are heat-treated to complete the eradication of solvent traces and then deodorized to remove unpleasant tastes. Defatted flour is made afterwards by grinding the flakes.

Some of the defatted flour goes directly into baby foods, biscuits, savoury snacks and health foods. The rest is the basis of Textured Vegetable Protein, or TVP. In a similar process to spaghetti manufacture, TVP is made by cooking the defatted flour in an extruder—a machine that looks like a giant mincer. Ribbons of TVP are extruded, sliced and dried. The nozzles on the extruder can be varied to produce chunks, granules, and other shapes and textures. The protein level of this TVP is around 50 per cent.

Spinning soya

Soya protein concentrate has a 70 per cent protein content when dry, and is obtained by removing the soluble carbohydrates from the defatted flakes. The flakes are treated with alcohol and gently heated. This 'fixes' the soluble protein. The resulting concentrate is then spray-dried to a fine powder and is used in the meat processing industry—small quantities are added to luncheon meat, frankfurters and sausages. The concentrate helps to bind the meat and stops the fats and juices from running out during cooking.

Soya protein isolate has a protein content of over 90 per cent when dry. The protein is extracted from the defatted flakes by treating them with a mild alkali solution that makes the protein soluble. The protein forms curds which are separated out, neutralized, washed, and dried to form a powder. The isolate

TVP manufacture has three distinct stages but each yields some edible soya products. The flakes gained at stage one are made into the functional ingredients of stage two—all four forms are added to food, ranging from bread to meat products, to increase protein and improve eating quality. Spun TVP costs most to make but can mimic other flavours best and has most protein.

THE THREE STAGES IN SOYA BEAN PROCESSING

is included in meat products, especially where they are subjected to high temperatures during processing. It is also used in bakery products, dairy foods, high-protein processed foods, special diets in hospitals and as a protein source in milk-free baby foods. It is also the starting point for the most sophisticated development in meat analogues, spun TVP.

Spinning is the process by which vegetable protein can be made to resemble real animal meats most perfectly. Firstly the soya protein isolate is returned to solubility, and its alkalinity increased to help it dissolve. The solution is then forced through tiny holes in a metal *spinneret*—a steel or platinum disc—and the thin jets of solution pass into a fixing bath containing a mild acetic acid solution where they coagulate immediately into thread-like fibres. Common salt in the acid solution draws the water out of the fibres.

Below A roadside café in China serving soya snacks. The soya bean is one of the five 'sacred' grains of ancient China.

The fibres are washed free of salt and acid, drawn out to up to four times their original length, and bound together with gum or a carbohydrate material such as starch, dextrin, alginate, or carboxy-methyl-cellulose.

The length of the stretching determines the 'chewiness' of the product. When the right point has been reached, the fibre bundles are flavoured, coloured, and cut and shaped according to the meat being simulated. Spun TVP is used extensively in institutional catering in mince form, and also as chunks with a variety of flavourings. Manufacturers claim that in a meat/vegetable protein mixture it is possible to use up to 50 per cent spun TVP without its taste being apparent.

Spun TVP is of course far more expensive than extruded TVP because of the greater number of stages involved. But it still represents a saving when mixed with animal protein, and it has a far superior 'eating quality' to extruded TVP.

New processes for accurately imitating real meat have seized the modern imagination and occupied much research. But this objective carries with it the danger that soya proteins will become an unachievable luxury in areas of the world where cheap proteins are most needed. The soya bean should be utilized both for richer nations which need to moderate the large amounts of uneconomically produced animal proteins currently dominating their diet and for poorer nations which simply need more protein. If current trends continue then by 1985 the average North American will be eating 1 kg of meat every three days and the average inhabitant of India 1 kg every 300 days.

One hope for the future lies in new strains of soya plants being developed for growth in all conditions. Already several African countries including Nigeria, Kenya, Zaïre, Zambia and South Africa have begun to grow soya. But the greatest successes have been in Mexico and India where strains from the USA have produced better yields than in their country of origin. Indeed, scientists forecast yields threefold the current levels and this may hasten the soya revolution.

Machine Technology: Transport

Below Millions of man-hours and vast sums of money have gone into the design and construction of the TGV and its track. But it is a sound investment—SNCF have won over thousands of former airline passengers.

France's high-speed train

Once upon a time trains were the fastest things on Earth not to be fired from the barrel of a gun; but with the rise in air travel and the spread of the motorcar, railways lost much of their public in the years after World War 2. Now they are fighting back: high-speed trains are among the most cost-effective means of transport over medium distances, and a whole family of these machines is being developed all over Europe.

Since September 1981 the world's highest regular railway speeds—260 km/h (161 mph)—have been run in France. A brand-new railway system designed to operate at these speeds has been built from Paris to Lyons, to supplement an existing route that was overloaded with traffic. Complete with its own special rolling stock, the new line cost some £1 billion ($1.8 billion), an expenditure justified by a forecast 26 per cent rise in French passenger traffic up to 1985.

The Paris-Lyons double-track line joins a family of high-speed European electric railways built to accommodate increasing traffic. Italy has opened the southern half of a line from Rome to Florence, West Germany is building several high-speed lines, while Spain is planning a Madrid-Barcelona railway. Outside Europe, Japan opened the pioneer high-speed route—the Shinkansen, from Tokyo to Osaka—in 1964.

All these railways (except those in Japan) were designed in the early 1970s for a maximum speed of 300 km/h (185 mph). However, when a number of European railway operators got together under the auspices of the International Union of Railways to examine the feasibility of high-speed rail travel, it was concluded that maximum speeds of over 270 km/h (167.5 mph) were not commercial—passengers simply would not pay to run any faster.

All the same, French National Railways (SNCF) tried in 1981 to find out just how fast their new train (called *Train à Grande Vitesse,* or TGV) could run. There was an element of 'cheating' in this—final drive gear ratios on the train selected were altered for higher speed, and the wheel diameters were increased by 130 mm (5 in.). In addition, the engine pulled a shorter and lighter train, and the voltage in the overhead wire from which power was drawn was increased.

The result of the alterations when a run was made on the new railway was a world rail speed record of 380 km/h (236 mph). Everything worked so well on this occasion that the French feel they might increase the speed limit for the railway if the extra revenue to be gained by a faster service can compensate for the higher running costs.

It is normal to expect that bigger locomotives will haul trains faster regardless of the existing installations. But at speeds of over 160 km/h (100 mph) a railway must be conceived as a fully integrated system. Track, electricity supply, motors, coaches and signalling must all be designed to complement one another, as well as being designed specifically for high speed.

The fast lane

It was decided very early that for the Paris-Lyons service the new line would be for high-speed trains only, slower trains and freight taking the old route. As a result, gradients of up to 1 in 28.5 (3.5 per cent) could be accepted on the new railway. Compared with old-style track restricted to 1 in 100 (1 per cent) gradients and tunnels, construction costs fell by 30 per cent, for a minimal increase in energy consumption during operation and a slight drop in average train speeds.

For high-speed running, track geometry must be precisely maintained—both horizontally and vertically—so no embankment settlement is acceptable, still less subsidence. On the TGV system each embankment is built up in compacted layers to give a firm track foundation. At the same time minimum curve radius on the railway is nearly 4 km (2.5 miles) and the angle at which the curves are banked is at the optimum for the TGV as there is no need to compromise for slower ones.

Two spurs connect with the old line—at Aisy near Dijon and at Pont de Veyle near Macon—and the points at these junctions have a speed limit as high as 220 km/h (136 mph) over the curved turnout. As the points are over 500 m (550 yards) long, the usual operating motor at the toe of the points is supplemented by additional motors along the tracks to ensure correct movement.

Railways south-east of Paris are electrified at 1,500 volts DC, but experience shows that single-phase electrification at 25,000 volts 50 Hz AC is more economical. For the first 30 km (18.5 miles) out of Paris the TGVs run on the old railway, but the 25,000 volt AC system has been adopted for the new line, with the result that all trains have to be equipped to take both overhead line voltages. Indeed, six train sets are equipped for a third Swiss supply voltage to permit a Paris-Lausanne service.

Another factor in the choice of 25,000 volt electrification is that it is particularly difficult for signal circuit breakers to distinguish between a 1,500 volt DC load and an outright short circuit over a long distance; on the old track the new trains are limited to 160 km/h (100 mph) for this very reason. Using a higher voltage with alternating current simplifies this design problem and also helps the trains on the track because it is able to employ much lighter electrical apparatus.

The train's electricity supply comes from a tensioned overhead wire by means of a flexible pantograph on its roof. The pantograph exerts constant pressure on the wire despite the train's lateral and vertical movements and the tendency for the wire to be thrown upwards as the train passes.

Six traction motors are slung under the coach bodies at each end of a train set, four under each locomotive unit and two each under the ends of adjacent coaches. Each motor drives the axle next to it through a

Below Despite the high speeds of which it is capable, TGV has only one driver. High speeds mean that signals must be displayed in the cab *(right)*, with contact between the driver and the signaller maintained by radio.

Machine Technology: Transport

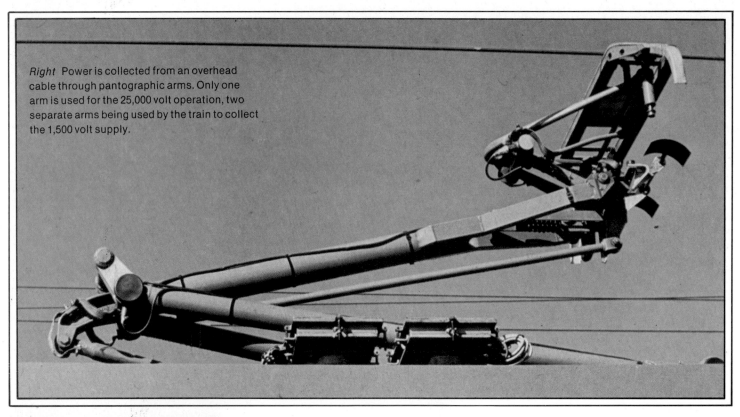

Right Power is collected from an overhead cable through pantographic arms. Only one arm is used for the 25,000 volt operation, two separate arms being used by the train to collect the 1,500 volt supply.

Above Design of the carriages and the gangway between them looks quite conventional in spite of the unusual bogie configuration. Hardware matters less to the passenger than comfort and reliability.

TGV—KEY TO CUTAWAY

The TGV is unconventional in many ways. An automatic control panel (1) governs the braking system. Immediately behind the cab (2) are cubicles containing electrical equipment (3) and motor circuitry (4). Below these are the cab's air conditioner (5) and storage batteries (6), while the main transformer (7) and control circuitry (8) lie amidships. 1,500 volt and 25,000 volt pantographs (9) collect the traction current; the 25,000 volt current is passed along the train through a roof-mounted cable (10). The motors themselves are mounted below the chassis (11) and drive one bogie on the power car and one on the adjacent carriage (12). Also in this first-class carriage is a transformer (13) which supplies power to the lights and pantries (14) throughout the train. Aerodynamic skirts cover the electrical equipment mounted below the carriages' chassis (15). These skirts control the airflow beneath the carriage.

TGV — HIGH SPEED TRAIN

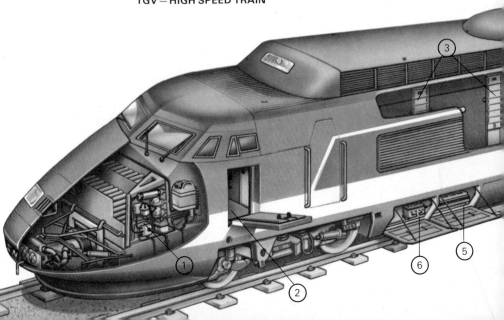

system of cardan shafts, and motor control is by voltage variation only.

Maximum tractive effort at the wheel rims for one train set is 6.35 mW, but only 4.2 mW are needed for 260 km/h (161 mph) running on the level. Rheostatic braking (where the motors act as generators instead of engines) can be fully applied at maximum speed, while at the same time clasp brakes are applied lightly to the wheel treads to clean them and prevent slip. Heat generated in the traction motors during braking is dispersed in air-cooled, roof-mounted resistors.

Below 200 km/h (125 mph), pneumatic disc brakes on trailing axles and the clasp brakes automatically blend in together while the rheostatic brake is taken out progressively. In an emergency, the train can stop in 3.1 km (1.86 miles).

The 87 new trains are made up into ten-coach sets, two of which can be run coupled together. Apart from the locomotive units at either end on two four-wheel bogies, the eight intermediate coaches are articulated; that is, the two adjoining coach ends rest on a common bogie, so for eight coaches only nine bogies are needed.

Effective springs and suspension are crucial in high-speed running. Of particular importance is the prevention of *hunting*—a rhythmic sideways movement of the bogie as it runs along—which can make the wheel flanges strike the rails with considerable force, so knocking them out of alignment. The trains are designed with a 17 tonne load on each axle, which compares with the usual 23 tonne axle load limit in France. Reduced axle load gives less wear and tear on the track at the higher running speeds.

Lightly sprung

The usual arrangements of springing between the axles and the bogie frame, supplemented by springs between the bogies and coach bodies, are used for the new trains.

Modern technology allows railway signals to be controlled from a central point—in this case Lyons. The familiar lineside signals are omitted, however, in favour of signals displayed electronically inside the cab. Signal interlockings set every 12 km (7.5 miles) are telecommanded from Lyons, as are the power sub-stations to feed the trains. Emergency telephones are sited every kilometre, as well as at all other strategic points, and train drivers can talk direct to the Lyons signalling centre by radio.

The guiding principle of electronic railway signalling is that only one train can be allowed to occupy one section of track at a time. The two rails of a track are insulated from each other and connected to a source of electricity, forming a track circuit. A train standing on the track joins the rails through its axles and completes the circuit. This then puts the signal behind it at 'danger'. The Paris-Lyons route is divided into track signal circuits, each 2.1 km long.

On most railways with single-phase AC electrification it is usual to have DC track signal circuits, but there are operational advantages in having AC signal circuits and various methods of achieving this have been tried on railways electrified at 25,000 volts AC. SNCF have evolved their own entirely new method for the Paris-Lyons railway.

High-frequency alternating currents are used to avoid interference to the signalling from the traction current (which is earthed through the tracks) and its possible harmonics. To prevent mutual interference bet-

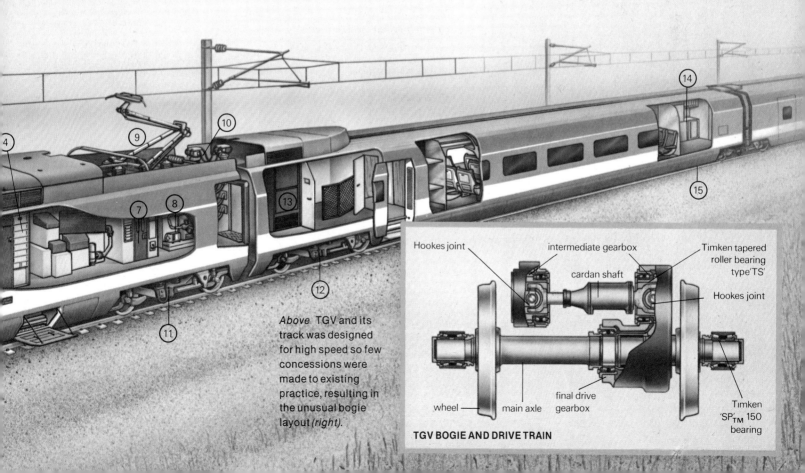

Above TGV and its track was designed for high speed so few concessions were made to existing practice, resulting in the unusual bogie layout (*right*).

TGV BOGIE AND DRIVE TRAIN

Machine Technology: Transport

ween signal circuits, frequencies switch from 1,700 to 2,300 Hz in one track section and 2,000 to 2,600 Hz in the adjacent one.

Without assistance, current at these frequencies will not flow more than 650-820 yards (600-750 m) through steel rails. Consequently the TGV signal circuit current is boosted slightly to flow the 2,100 m of a section and no more, thus avoiding insulated rail joints at the end of each section. This in turn reduces cost and helps the civil engineers who prefer to weld their rails together in the longest lengths possible as joints can cause trouble and be expensive.

The frequency of the track circuit currents are modulated between plus and minus 10 Hz, the rate of modulation or pulsing giving 15 possible pulse rate codes. Ten codes are in use between 618 pulses/min and 1,740 pulses/min, together with an eleventh given when there is no pulsing or no current. The codes are picked up by the train through induction windings mounted at the front just above the rails. On board the codes are amplified, decoded, and operate a cab signal.

Cab signals

One of the cab signals shows the speed at which the train should be running at the end of a section in km/h. When all is well the driver sees the letters VL (standing for *vitesse limité*) meaning the line maximum speed of 260 km/h. When a train needs to be stopped the signalman at Lyons causes the interlockings to get up the necessary track circuits. The first track circuit encountered by the train puts up the figures 260. The next two track circuits show the figures 220 and 160, while the final track circuit puts up 000 on a red background for the stop.

A post carrying a reflecting panel is placed at the end of each track circuit, which gives the driver a mark at which to stop. Marker posts protecting pairs of points over which there may be conflicting train movements display a red light in addition when a stop is necessary. An ordinary service stop uses four 2.1 km track circuit sections—a total distance of nearly 8.4 km (5.25 miles).

Below More speed, less airline traffic. TGV's efficiency has given the airlines a few problems! The unusual bogie and suspension layout *(left)* cuts down weight and improves high-speed stability.

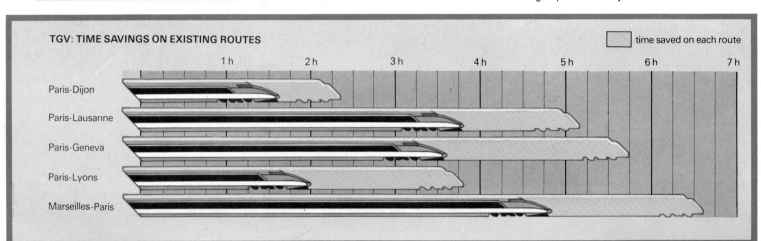

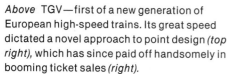

Above TGV—first of a new generation of European high-speed trains. Its great speed dictated a novel approach to point design *(top right),* which has since paid off handsomely in booming ticket sales *(right).*

Europe's first fully automatic passenger railway is scheduled to open at Lille during 1982, the trains carrying no railwaymen on board. The technology exists to automate the Paris-Lyons line as well, but automation cannot take place as long as sections of track in the Paris area and at Lyons are shared with conventional rolling stock operating on the old-style system.

Currently the new service from Paris to Lyons offers 15 trains a day. Three of the trains are extended beyond Lyons to St Etienne, and there are also two Paris-Geneva services that use the new line only in part. When the new railway is fully open in 1983 the Paris-Lyons time will be cut to 2 hours —an average speed of 213 km/h (132 mph).

At 43 kW/h per seat per kilometre, the TGV's energy consumption will be only slightly over that of previous express trains. Railwaymen point out with satisfaction that this energy consumption figure is doubled by the motorcar and multiplied by four in the case of an aircraft.

European railways have always competed successfully with air services, but the new high-speed routes will increase the distance over which such competition is possible to some 650 km (400 miles). For one-day business journeys the competitive distance will be about 400 km (250 miles).

Without doubt, modern express trains are comfortable, fairly convenient, and comparatively cheap; but they have been handicapped by the time that some journeys take. Passengers have long wanted the benefits of high-speed air travel at an affordable price, and this is a challenge which is at last being met—at least in part—by new developments like the TGV.

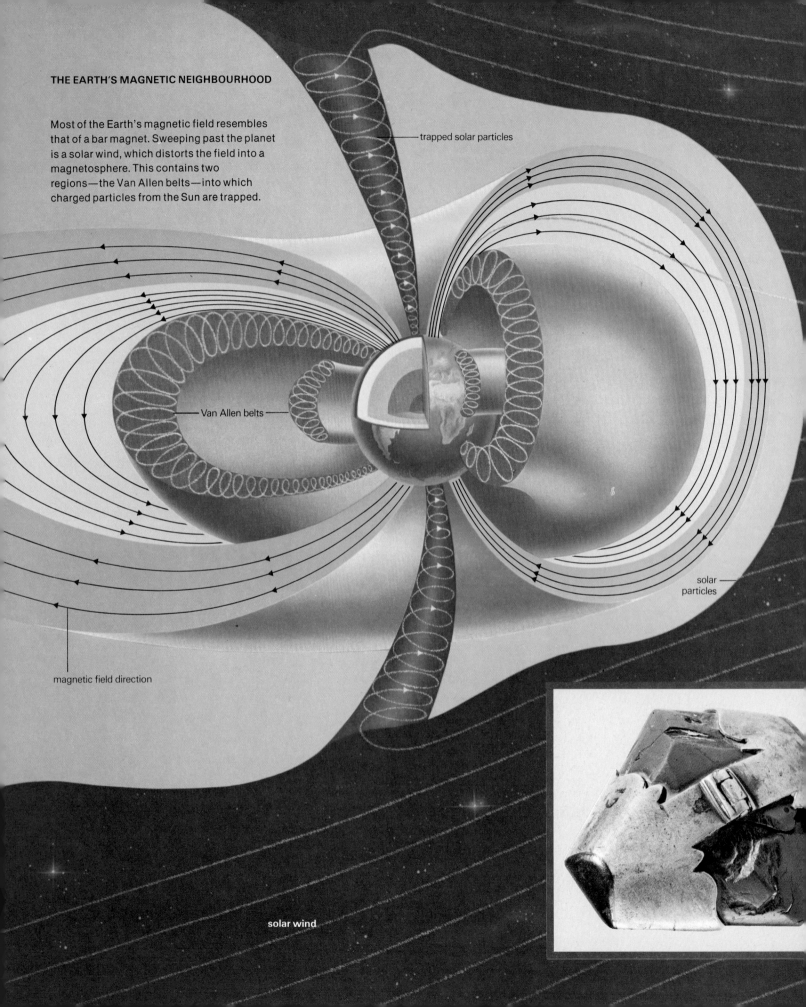

THE EARTH'S MAGNETIC NEIGHBOURHOOD

Most of the Earth's magnetic field resembles that of a bar magnet. Sweeping past the planet is a solar wind, which distorts the field into a magnetosphere. This contains two regions—the Van Allen belts—into which charged particles from the Sun are trapped.

trapped solar particles

Van Allen belts

solar particles

magnetic field direction

solar wind

Frontiers: Earth Science

Earth's magnetic personality

Throughout the centuries, mariners navigated the globe by noting the polarity of the Earth's magnetism. Now scientists have shown that magnetism also helps some animals to find their way over large distances and that the same phenomenon might even have an affect on the world's climate.

The idea that there might be a link between *geomagnetism* (the Earth's magnetism) and climate is at least 30 years old, but only recently has it been argued convincingly. During the past 10,000 years, and possibly for several hundred thousand years, it appears that the geomagnetic field was weaker while the temperature at the Earth's surface was generally warmer. At first this observation might appear to have no significance because the depths of the Earth's interior—where magnetism originates—has no link with the atmosphere. On closer examination, however, it seems likely that magnetism and climate are linked indirectly. The temperature of the atmosphere is modulated by changes in the tilt of the Earth and the shape of the planet's orbit round the Sun, and such changes in the Earth's motion are also likely to affect the motions of the fluid in the Earth's core.

But if the connection between magnetism and climate is indirect, that between magnetism and some forms of animal behaviour may be closer. Biologists have long suspected, for example, that pigeons use the Earth's magnetic field as part of an elaborate guidance system, but the hypothesis has been difficult to prove. In recent years, however, pigeons have been found to have minute particles of magnetic material in their head, which suggests, if it does not prove, that the idea of magnetic guidance might be correct.

Likewise, small magnetic particles have been found in the head of some dolphins—creatures that also seem to know just where they are going. This has led to the suggestion that dolphins could be deflected from accidental capture by ensuring that fishing nets carry magnetic devices.

Strong fields

Apparently, bees are also affected by geomagnetic fields. When the strength of the magnetic field around bees is increased to about ten times the strength of the Earth's field, the bees no longer perform their 'dance'. This suggests strongly that the dance is governed by a magnetic mechanism which breaks down in a strong field.

The Earth's magnetism is known by its effects. A magnetic compass needle free to rotate on a vertical pivot will come to rest pointing towards north. But at most places on the Earth's surface, the compass needle does not point exactly towards geographic north. There are places, even in middle-to-low latitudes (between 45°N and 45°S), where the compass points more than 40° away from true north; and at higher latitudes, the angle between the needle and geographic north (known as the *declination*) is often far greater. Moreover, the declination is changing continuously, so nowhere does the compass needle point in exactly the same direction in any two consecutive years.

The bulk of the present magnetic field is simple; but superimposed upon this main field are many small, highly irregular, fields that produce an overall impression of great complexity. More than 90 per cent of the magnetic field is that of a *dipole,* the simplest magnetic system known to exist. It is identical in shape to that of a bar magnet, which has two poles—a north pole at one end and a south pole at the other. This does not mean, however, that the Earth actually has a bar magnet at its centre: the planet merely behaves as if it has.

The conventional magnetic compass can react only to that part of the Earth's field acting in a horizontal direction, because the needle is constrained to move only in the horizontal plane. But if there were such a thing as a compass needle free to rotate in any possible direction (that is, both vertically and horizontally at once), it could be used to plot the field around the Earth in three dimensions because a free compass needle will always come to rest along the field.

Such an exercise would demonstrate the essential simplicity of the geomagnetic field, but it would also reveal other important characteristics of the field. For example, although the field pattern is symmetrical about the Earth's centre, it is not symmetrical about the axis of rotation. The dipole axis slopes at 11° to the rotational axis. The geomagnetic poles—the points at which the dipole axis cuts the Earth's surface—do not, therefore, coincide with the geographic poles; they lie 11° of latitude away. The North Geomagnetic Pole is at 79°N, 70°W and the South Geomagnetic Pole is at 79°S, 110°E.

Opposite sense

It is interesting to note that the bar magnet equivalent at the Earth's centre is the opposite way round. The south pole of the bar magnet equivalent (or of the dipole) lies in the Northern Hemisphere and corresponds with the Earth's North Geomagnetic Pole. This must be so if a compass needle points north because a north and a south pole attract each other, whereas north poles (or south poles) repel each other. Thus the north

LODESTONE IN BRASS

By at least 600 BC, the Greeks were familiar with the strongly magnetic rock known as lodestone. In the 17th century, the material was mounted in a brass holder and used to induce magnetism in compass needles by stroking one needle with the other.

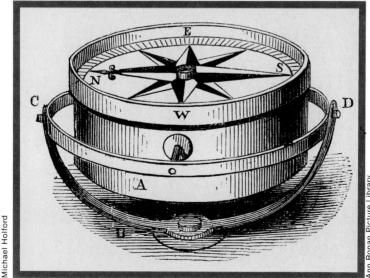

Left The needle of a mariner's compass comes to rest in the direction of the geomagnetic field, the influence of which was observed long before the field itself was discovered.

Left Weather on Earth is much affected by atmospheric temperature—itself modulated by changes in the planet's orbital motion, which also affects the fluid magnetic core.

pole of a compass needle points towards the south pole of the dipole.

The 11° slope between the dipole axis of rotation helps to explain why compass needles do not generally point true north. They point towards the North Geomagnetic Pole, and will only point towards the North Geographic Pole as well if the compass, the geomagnetic pole and geographic pole all happen to be exactly in line.

But theoretically, a compass needle should never point more than 11° away from the geographic pole, because the geomagnetic and geographic poles are only 11° apart. Why then, is it possible to have declinations of 40° or more? The reason is that the geomagnetic field is not just that of a simple dipole. It is not even just that of a simple dipole inclined at 11° to the rotational axis. It is that of a simple inclined dipole with many weaker fields superimposed randomly.

One way of thinking about this is to imagine a football with a strong bar magnet fixed at its centre. Then imagine that the space between the central magnet and the football's shell is filled with hundreds of smaller and very much weaker bar magnets arranged in all directions. The sort of field that would be observed outside the football is largely that of a simple dipole, but highly distorted by the overall effect of the jumble of small magnets.

The irregular fields all together are known as the *non-dipole field*. Its average strength over the Earth's surface is less than ten per cent of that of the dipole field, although at particular locations it can amount to as much as 30-40 per cent. The combined effect of the non-dipole field and the 11° inclined dipole can, therefore, produce large declinations at certain places.

The non-dipole field has another important effect. If the geomagnetic field were solely dipolar, a free compass needle would point downwards vertically at the North Geomagnetic Pole and upwards vertically at the South Geomagnetic Pole. In fact it does neither, because at both places there are non-dipole fields also deflecting the needle.

There are places elsewhere, however, at which the particular combinations of dipole and non-dipole fields cause a needle to point vertically. They are called the North and South Magnetic Dip Poles. To see just what the non-dipole field itself is like, it is possible

to subtract the dipole field from the total field and then plot the resulting non-dipole field as a contour map. This reveals the irregularity of the non-dipole field, but only at a particular moment.

Systematic recording of the geomagnetic field direction began only about 150 years ago, but the earlier of these observations were neither complete nor accurate. Nevertheless, there are enough data available to enable scientists to detect the chief short-term changes in the field. The most rapid changes occur in the irregular non-dipole field, which appears to be drifting westwards at a rate of about 0.2° longitude a year. This appears to imply that the whole non-dipole field should move completely around the Earth in about 1,800 years, at the end of which it would look much the same as when the cyclic movement began.

Constant changes

In fact, the closed loops in the non-dipole field chart do not persist indefinitely in the same form. They grow and decay, expand and contract, and disappear and reappear on time scales of 100-1,000 years, so that after much less than 1,800 years the non-dipole field has changed its shape completely.

The dipole itself does not stay entirely still either, although it moves much more slowly than its non-dipole counterpart. During the past 150 years, the geomagnetic poles have remained at latitude 79° but have moved through about 6° of longitude.

It is now known that the angular separation of the two types of pole has not always been 11°. This knowledge comes from the

Despite the Earth's solid appearance, there is evidence from volcanoes *(right)* of a molten core. But even the rocky crust wells up in currents much like boiling water in slow motion. Much of this material is magnetic, and scientists explain the origin of Earth's magnetism by the principle of a self-exciting dynamo *(below)*. In this, a disc is kept rotating by the effects of its own motion on an induced current and induced magnetism.

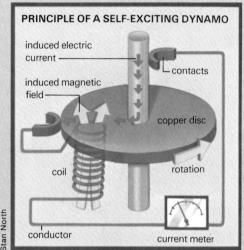

record in rocks, from which it is now possible to learn what the geomagnetic field was like many millions of years ago.

Most rocks contain small amounts of magnetic compounds, usually oxides of iron and titanium. When a rock forms, the magnetic material in it becomes magnetized weakly along the direction of the prevailing field. Moreover, the magnetization is usually stable, being retained in the rock indefinitely, so the rock acts as a very accurate geomagnetic field recorder. By measuring its magnetization on extremely sensitive equipment, experts can deduce the position of the

pole at the time the rock was laid down.

When the geomagnetic poles deduced from rocks up to, say, 10,000 years ago are plotted on the globe, they are grouped not around the present geomagnetic pole but around the geographic pole. Indeed, if the rock-derived poles are put together to make an average pole, that average is almost exactly the geographic pole. This is a most significant result, because it shows that when averaged over a sufficient period of time the Earth's dipole lies along the axis of rotation.

Scientists believe that the dipole wobbles about randomly within strict limits, so that the geomagnetic pole wanders around in the vicinity of the geographic pole. The two poles coincide on average, but at any given time they can be separated by anything up to

Left The seasonal migration of animals over huge distances is remarkable for its precision. Recent research links this navigational feat with a magnetic sense.

Frontiers: Earth Science

10°-15°. For the past 150 years, the separation has been 11° but this particular angle has no other significance.

A much more spectacular, if less expected, phenomenon to be discovered from the magnetism of rocks was field reversal. There is now irrefutable evidence that at various times in the past the dipole was not in its present (normal) direction, with its south pole in the Northern Hemisphere, but in precisely the opposite (reversed) direction, with its north pole in the Northern Hemisphere. When the dipole (and hence the main geomagnetic field) was reversed, a compass needle would point not northwards, as it does now, but southwards.

The process of reversal probably takes about 10,000 years. The dipole does not just flip over, however; its strength first decreases to zero and then builds up again but with the dipole in the opposite direction. In fact, the reversal process can be seen as part of a much broader pattern of variation in the strength of the dipole. Apparently, the

The extensive use of radio and radar for navigation at sea has not replaced the mariner's compass on the bridge *(centre)*. Radar is invaluable for fixing a ship's position, but once a course has been decided the ship is steered by compass. One modern compass *(right)* has a moving dial—instead of a needle—which is divided into 360 degrees and labelled with the major directions to aid easy reference.

CHANGES IN STRENGTH AND DIRECTION OF THE EARTH'S MAGNETIC FIELD

The Earth's dipole varies randomly in strength and direction.

dipole fluctuates continuously in strength but only comparatively rarely does a decrease in strength lead to a full reversal.

Using satellites and space probes, scientists have studied the Earth's magnetic field from space. The field extends huge distances into space but is distorted in places by the influence of the Sun, against which it is a useful filter from harmful radiation.

The geomagnetic dipole strength is on a downwards trend now and has been so for about 2,000 years. For the past 150 years, in particular, it has been decreasing by about five per cent a century; and recent observations, both at the Earth's surface and from the MAGSAT satellite, have confirmed that the decrease is continuing. It is impossible to say whether the poles will reverse in about 2,000 years, because the field strength is just as likely to start increasing again as it is to continue its descent to an actual reversal.

Soon after the existence of field reversals had been proved during the 1960s, and it had become clear that during a reversal the strength of the geomagnetic dipole decreases

to zero, some scientists began to look at the implications for other terrestrial phenomena. One suggestion, for example, was that field reversals might explain the numerous extinctions of species during the history of life on Earth. The argument was that once the dipole had disappeared, there would temporarily be no magnetic field barrier against cosmic rays, which would then pour down on to the Earth's surface and cause genetic damage to plants and animals.

Unfortunately, this particular idea has not been supported by the evidence. Although one or two extinction episodes coincide with known field reversals, there have been many more extinctions without reversals and reversals without extinctions. Moreover, when the dipole field decreases to zero during a reversal, the non-dipole field generally does not. A magnetic barrier against cosmic rays thus remains, even during a reversal.

The simplest explanation of the Earth's field—first suggested by the London physician William Gilbert in 1600—is that the Earth is a huge, solid magnet. But there are many reasons why this cannot be so, the most crucial of which relates to the rapidity with which the field changes. If any part of the Earth's solid interior were to move at the speed implied by the observed field changes, the planet would long ago have disintegrated under the strain.

The rapid field changes also provide a clue, however, because rapid motion implies a source in a fluid. The only place in which the magnetic field could possibly originate, therefore, is the Earth's outer core which consists largely of molten iron.

How the core generates the field is uncertain. The only mechanism suggested so far is the *self-exciting dynamo*. This is based on two fundamental laws of physics—that when an electrical conductor moves in a magnetic field an electric current is induced in it and, conversely, that a current flowing in a conductor generates a magnetic field.

The electrically conducting iron in the Earth's outer core is in constant motion. Assuming the presence of a small magnetic field in the beginning (it could have been a stray field from the Sun), the molten iron would move in that field and generate a current. This current would then flow in the iron, producing more magnetic field through which the iron would move to generate more current, and so on.

The details of how the self-exciting dynamo works will probably remain buried deep in the Earth. But the principle is valid because it has proved possible to built a working model of a simple self-exciting dynamo, using wires and rotating discs or cylinders. Perhaps the greatest puzzle about the Earth's magnetic field is what influence it exerts on our lives. We have some clues, but the full truth may never be known.

The origin of geomagnetism might be explained by the principle of a dynamo that powers itself. Such a system has been demonstrated on a prototype *(below)* made from motors.

Medical Science: Intensive Care

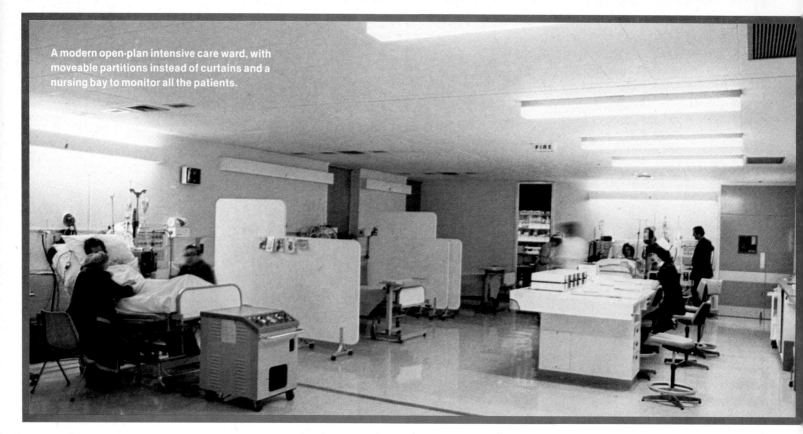

A modern open-plan intensive care ward, with moveable partitions instead of curtains and a nursing bay to monitor all the patients.

The life supporters

Few branches of medicine owe as much to the development and application of new technology, or have advanced so rapidly within such a short space of time, as intensive care. Techniques which a new generation of doctors now take almost for granted include mechanical inflation of the lungs, rapid measurement of oxygen levels in arterial blood to optimize control of the artificial ventilation, and the use of tracheostomy to secure a clear and airtight connection between the lungs and an external ventilator.

The hospital's intensive care unit has become the focal point of a multidisciplinary team which combines the skills of anaesthetists, physicians, surgeons, nurses, biochemists and engineers. It is now larger and more complex and has at its disposal resources of much greater technological sophistication than ever before.

Almost by definition, all of the patients in an intensive care unit are critically ill, but whether the cause is injury, disease or major surgery, many of the problems encountered in their care are the same.

The majority of patients admitted to intensive care units have respiratory problems which are serious enough to warrant artificial ventilation. Since breathing is controlled by special centres in the brain and breathing movements depend upon the function of respiratory muscles and the nerves which control them—as well as the integrity of the chest wall—brain damage, nerve and muscle disorders or severe injury to the chest wall might all contribute to respiratory failure. So, too, may many different kinds of lung diseases and injuries.

The first mechanical ventilators were huge, cumbersome devices. Some, like the old 'Iron Lung', actually enclosed the patient, inflating and deflating the lungs by altering the pressure outside the chest wall.

Today, ventilators are compact and impressively sophisticated. Many incorporate microprocessors, affording precise control of the proportions of oxygen, nitrogen and carbon dioxide delivered to the patient. The gas mixture can be humidified and warmed to the body temperature, and drugs can be added—to relax the air passages, for example.

Time, pressure and flow control the depth of each breath, and in many machines each may be given priority in much the same way as an automatic-exposure camera offers the photographer a choice between aperture or shutter-speed priority.

Cycling is the term used to describe the mechanism by which the inspiratory phase of each breath (the rate of flow throughout the inspiration) is terminated. In time cycling, the duration of each inspiration is merely pre-set. In pressure cycling, inspiratory flow continues until a pre-determined pressure is reached. In volume cycling, flow is generated by a mechanically driven piston, set to deliver a pre-selected volume of gas.

Patient-triggered ventilation

Patients undergoing ventilation are often unconscious as a result of their illness; if not, it is often easier to ventilate them if they are sedated and given muscle relaxants. But if they are conscious or semi-conscious, it may be preferable in some situations for patient-triggered ventilation to take place.

An adjustable sensor in the ventilator responds immediately to the first sign of respiratory effort, completing the breath mechanically. There is a facility called *intermittent mandatory ventilation* for pre-setting the ventilator to deliver breaths at a

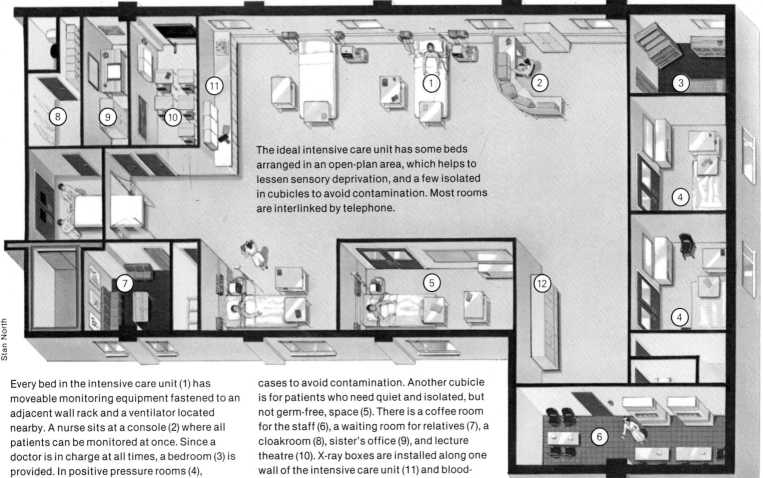

The ideal intensive care unit has some beds arranged in an open-plan area, which helps to lessen sensory deprivation, and a few isolated in cubicles to avoid contamination. Most rooms are interlinked by telephone.

GROUND PLAN OF AN INTENSIVE CARE UNIT

Every bed in the intensive care unit (1) has moveable monitoring equipment fastened to an adjacent wall rack and a ventilator located nearby. A nurse sits at a console (2) where all patients can be monitored at once. Since a doctor is in charge at all times, a bedroom (3) is provided. In positive pressure rooms (4), pressure blows air outwards only—preventing germs from entering and enabling isolation cases to avoid contamination. Another cubicle is for patients who need quiet and isolated, but not germ-free, space (5). There is a coffee room for the staff (6), a waiting room for relatives (7), a cloakroom (8), sister's office (9), and lecture theatre (10). X-ray boxes are installed along one wall of the intensive care unit (11) and blood-taking equipment is kept in cupboards arranged along another wall (12).

minimum frequency in case patient effort is not forthcoming.

The expiratory phase of each breath is usually passive, relying upon the natural elastic recoil of the lungs. Negative pressure is dangerous and is not used. It has been found, however, that impeding the outflow of expired air usually results in increased absorption of oxygen by the lungs. This impedence is known as PEEP (positive end expiratory pressure), and the ventilator's PEEP control varies the pressure of expired air.

The ventilator is connected to an endotracheal tube, which has an inflatable collar at its tip, making an airtight connection with the lungs. Such tubes are initially inserted via the mouth or nose, but if ventilation exceeds 48 hours, a tracheostomy is more comfortable and much safer.

Patients on ventilators require frequent physiotherapy—as well as suction with a fine catheter passed down the endotracheal tube to remove mucus and secretions that might cause obstruction or facilitate infection. The condition of the lungs is monitored by frequent examination and daily chest X-rays to enable complications to be detected early on.

Ventilation is monitored by measuring arterial blood levels of oxygen and carbon dioxide, as well as blood acidity (which partly depends on the ability of the lungs to remove carbon dioxide from the blood). An arterial blood sample is usually taken from the wrist. Many intensive care units now have microprocessor-controlled auto-analyzers in a special mini-laboratory adjoining the intensive care suite. They are simple to operate, and give an instant print-out.

Although artificial ventilation of the lungs is now a commonplace procedure, it is not yet possible to take over the function of the heart by artificial means in quite the same way. It is only feasible to maintain artificially powered circulation for short periods of time—during cardiac surgery, for example.

However, electronic monitoring of cardiac activity is undertaken routinely on most patients in intensive care units. Three adhesive electrodes on the patient's chest link up to a continuous video display of the electrocardiograph—which can be observed at the bedside as well as remotely in the nurses' office. This enables immediate action to be taken in the event of cardiac arrest, and also facilitates early diagnosis of abnormalities of heart rate and rhythm. There is usually a simultaneous digital display of the patient's pulse rate.

Blood pressure is another reflection of cardiac function, and can be monitored electronically by a tiny electronic pressure transducer, a device which can be inserted into a small artery in the wrist or foot. The venous pressure, also important, is usually measured by inserting a fine plastic catheter into a vein in the arm or neck and passing its tip into one of the large central veins leading to the heart. A rise in venous pressure may indicate cardiac failure, as may a fall in arterial pressure.

A more accurate index of cardiac function, however, is the 'wedge pressure' in the pulmonary artery (the main artery from the heart to the lungs). This is measured by passing a long, fine catheter through an arm or neck vein, into the right atrium and ventricle and on into a branch of the pulmonary artery. A tiny balloon is inflated close to the tip of the catheter. The pressure at the tip of the catheter, beyond the balloon, gives useful information about the function of the heart.

In cardiac failure, pressure in the blood vessels of the lungs builds up and fluid accumulates—contributing to respiratory failure. The wedge-pressure technique enables

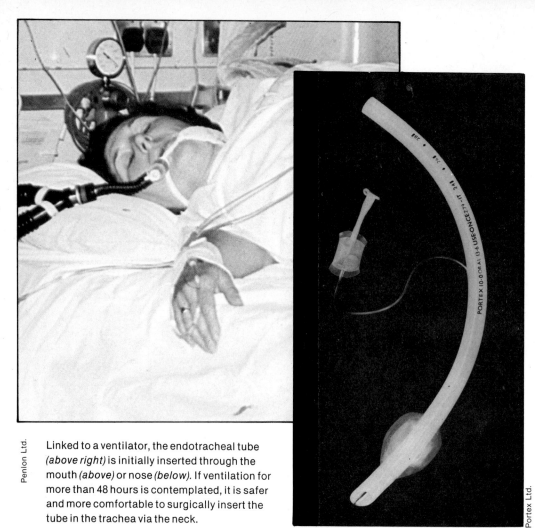

Linked to a ventilator, the endotracheal tube *(above right)* is initially inserted through the mouth *(above)* or nose *(below)*. If ventilation for more than 48 hours is contemplated, it is safer and more comfortable to surgically insert the tube in the trachea via the neck.

treatment to be given at a very early stage.

Blood pressure and venous pressure also reflect blood volume and fluid balance. Restoring central venous pressure to normal following sudden and severe blood loss indicates, for example, that the correct volume of blood has been transfused. Too much circulating fluid overloads the heart and contributes to cardiac and respiratory failure. Inadequate replacement reduces the oxygen supply to tissues and organs, causing shock.

Unconscious patients in intensive care units, unable to drink, lose fluid and salts in urine and perspiration. It is usual to compensate for these losses by administering fluids intravenously. Urine output is carefully monitored by means of a catheter in the bladder, and fluid and electrolyte replacement is carefully calculated. Frequent checks are made on blood levels of sodium, potassium, urea and bicarbonate.

Supplying nutrients

Great advances have been made in the field of nutrition of critically ill patients. Intravenous glucose is used as an energy source initially, but if a long stay in intensive care is anticipated, other measures are taken since intravenous feeding carries the risk of infecting and irritating blood vessels.

If the intestinal tract is undamaged, a fine feeding tube may be passed via the nose or mouth into the stomach and specially formulated feeds administered. In some situations it may even be desirable to perform a gastrostomy—a surgical opening through the abdominal wall and into the stomach, into which nutrients may be given.

Acute kidney failure and abnormalities in kidney function are surprisingly common problems in critically ill patients. Sudden and considerable blood loss—caused, for example, by a severe injury—reduces blood flow to the kidneys below a critical level.

The kidneys normally have a rich blood supply, receiving one quarter of the output from the heart. They tolerate this reduction in flow badly, and cells in the renal tubules (the microscopic functional sub-units of the kidney) are killed.

This condition may occur in a wide variety of situations—following surgery, burns, haemorrhage, a severe crush injury, massive infection or poisoning. Should the patient survive the effects of the underlying cause as well as the effects of kidney failure, however, kidney function usually returns.

In the intensive care unit, it is usual to screen frequently for early signs of kidney failure using blood tests and urine analysis.

Medical Science: Intensive Care

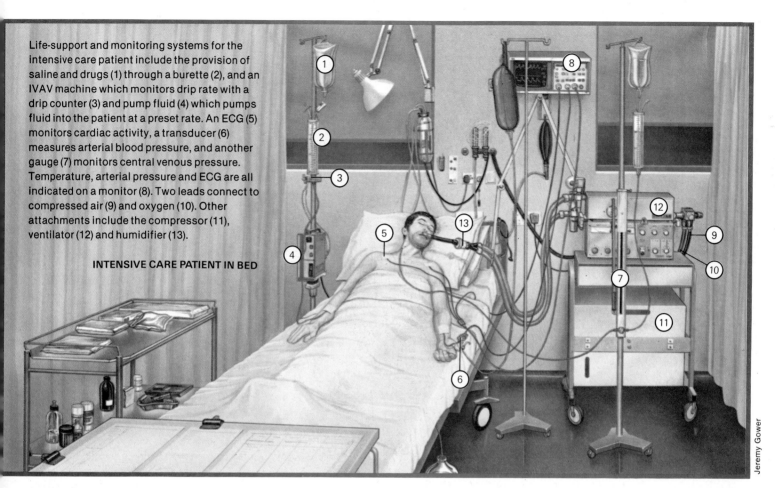

Life-support and monitoring systems for the intensive care patient include the provision of saline and drugs (1) through a burette (2), and an IVAV machine which monitors drip rate with a drip counter (3) and pump fluid (4) which pumps fluid into the patient at a preset rate. An ECG (5) monitors cardiac activity, a transducer (6) measures arterial blood pressure, and another gauge (7) monitors central venous pressure. Temperature, arterial pressure and ECG are all indicated on a monitor (8). Two leads connect to compressed air (9) and oxygen (10). Other attachments include the compressor (11), ventilator (12) and humidifier (13).

INTENSIVE CARE PATIENT IN BED

Severe cases are treated with dialysis. Dialysis brings blood into close contact with a specifically formulated dialysis fluid — separated by only a semi-permeable membrane which does not permit diffusion of substances of large molecular size but through which small molecules such as water, salts and toxic body waste products can pass freely.

In haemodialysis, blood from the patient flows over a cellophane membrane with a large surface area (usually about 2.5 sq m), and exchange takes place with dialysis fluid before the blood passes back from the dialysis machine to re-enter the patient's circulation.

Peritoneal dialysis is much simpler to manage, however, and is now a commonplace procedure in most intensive care units. The membrane is internal instead of external. In fact, it is the outer lining of the abdominal contents and intestines—the peritoneum. A catheter is passed through the abdominal wall into the peritoneal cavity, and dialysis fluid is allowed to enter and then drain from the abdomen at controlled rates. The composition of the fluid selected determines precisely how much fluid or salts enter or leave the bloodstream.

Dialysis is also sometimes used in severe cases of poisoning when there is no other means to remove toxic substances rapidly from the body. However, a new technique is also now available—*haemoperfusion*. Activated charcoal has a great capacity for binding and absorbing various substances. Haemoperfusion involves the passage of blood through a cartridge containing cellulose or albumin-coated charcoal bound to a film or mesh. The system is simpler to operate than haemodialysis, and more effective in many situations than peritoneal dialysis. Blood is simply pumped through the disposable or rechargeable cartridge and back into the patient.

Accidents and injuries

Accidents and injuries are the major cause of death in the under-30 age group, and they are responsible for up to one third of all hospital admissions. Improvements in intensive care are changing the outlook for severely injured patients.

Ventilation and monitoring for coma and other common complications of head injuries is essential. In some units, intracranial pressure monitoring is used as a guide to treatment. A tiny electronic measuring device, inserted through a small hole drilled into the skull, gives a digital display of intracranial pressure. Increases in pressure reflect swelling of the brain—a common sequel to injury—as well as intracranial bleeding. Early detection of change allows prompt treatment to be given.

Chest injuries may result in injury to the heart, lungs or major blood vessels. If so, immediate surgery is necessary—as well as adequate replacement of blood loss. A penetrating wound to the chest wall allows air or blood to enter the space between lung and chest wall, collapsing the lung. Special treatment to drain air or blood from this space is necessary for re-inflation of the lung.

Damage to the rib cage with multiple fractures produces a 'flail' chest. The respiratory muscles are unable to draw air into the lungs because the chest wall has lost its rigidity. The damaged flail segment of chest wall is mobile, moving inwards with inspiration and outwards with expiration. Artificial ventilation is essential.

Major burns are usually treated in specialized regional burns units. The main problems include fluid loss through the

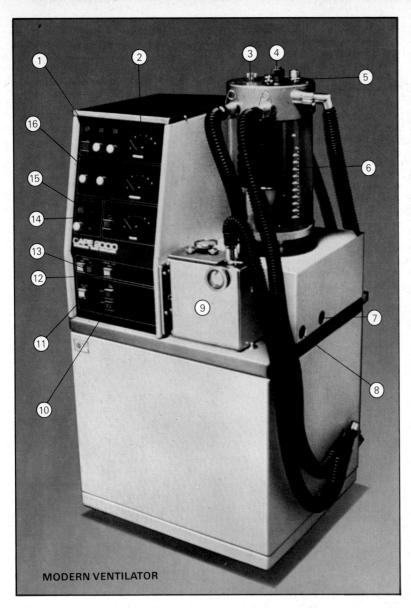

MODERN VENTILATOR

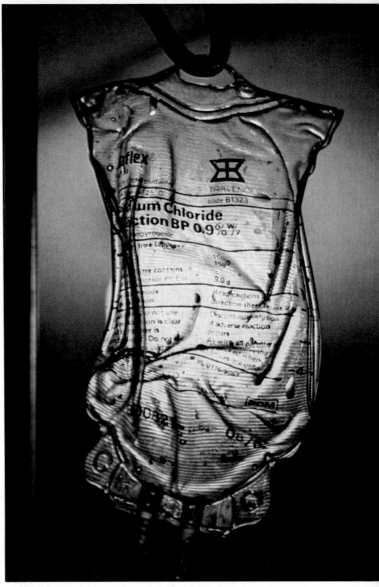

Above left A modern ventilator, showing (1) failure alarm, (2) pressure monitor, (3) inspiratory flow control, (4) PEEP valve, (5) manual positive pressure valve, (6) ventilating head, (7) air/oxygen inlet, (8) expiratory outlet, (9) humidifier, (10) air volume controls, (11) on/off switch, (12) expiratory filter controls, (13) humidifier control/warning light, (14) expiratory flow monitor, (15) intermittent ventilation control and (16) inspiratory/expiratory ratio controls. *Above right* A saline drip bag used to administer fluids and drugs intravenously.

damaged areas resulting in shock; lung injuries from smoke, heat, and fumes; kidney failure from the massive overload of waste products which are produced; and infection.

Advances in surgery—especially neuro- and cardiac surgery—have made it increasingly necessary for many patients to receive intensive care post-operatively so that their recovery can be carefully monitored. Prolonged ventilation may be required, as well as specialized nursing procedures.

Intensive care nursing requires continuous day and night care, with never less than one nurse allocated to each patient. The unconscious patient, unable to move, must be turned at least every two hours to prevent the development of pressure sores. The eyes, nose, mouth and skin must all be cleaned and cared for. In conscious patients, the psychological and emotional problems of coping with the strange and often frightening environment of the intensive care unit demand sympathy, understanding and skill at communication. The nurse has also the daunting task of looking after an expanding battery of electronic and mechanical devices monitoring every accessible bodily function, and responding to any abnormalities.

The doctor in overall charge of the intensive care unit is usually an anaesthetist or a general physician. But the key to the success of the unit is teamwork, which depends upon the joint efforts not only of doctors and nurses but also of laboratory staff, physiotherapists, technicians, engineers and the X-ray department.

In a setting where life support, ventilation and monitoring are the foremost priorities —and even a routine—it can become remarkably easy to devote less attention to the pressing need to investigate, diagnose and treat underlying problems. This sometimes happens in the case of patients with severe head injuries, for instance —partly because specialist neurosurgical skills and facilities are relatively scarce. Moreover, the maze of tubes, wires, digital read-outs and machinery can make it easy to lose sight of the patient as an individual.

In elderly patients, the terminally ill, the chronically sick and those for whom hope has faded, intensive care prolongs dying and too often destroys the dignity of death. A supreme challenge in intensive care medicine is simply to know when efforts to sustain life should cease, or not be made.

Machine Technology: Metallurgy

Forging ahead

More than ever, industry depends on the techniques of working hot metal from which essential machinery is made. One process—metal forging—is the craft of the blacksmith brought up to date. Aided by microprocessors, today's forge master can work aircraft turbine blades to an accuracy of a few hundredths of a millimetre or process tough, heat-resistant metals into thousands of identical components in only an hour's work.

Many complicated shapes can be formed from solid metal, using processes such as turning and grinding. But for certain critical components, shaping the metal by forging not only simplifies manufacture but also aligns the metal's crystalline structure—so making it extremely strong in required directions. In this way, a variety of different components can be developed with optimum metallurgical qualities.

The starting point in the manufacture of a forged component is an *ingot,* cast from molten metal. In ingot form, the metal contains many small cavities which, together with a lack of grain alignment, makes it very

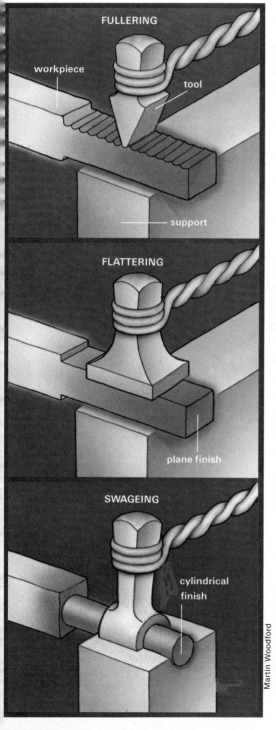

Left and below The techniques for forging large, unwieldy components are much the same as those used by a blacksmith, but the equipment is on a much grander scale.

VERTICAL FORGING PROCESS

- punch
- billet
- side die

side dies close and vertical ram completes forging

dies opened

finished forging

The manufacture of a gear cluster *(below)* begins with a single shaft, the diameter of which is increased in various stages *(right)*.

brittle and also structurally weak in places.

To form the ingot into, say, a liner's propeller shaft or any other large but critical component, it is squeezed between the jaws or tools of a powerful hydraulic press, some of which can handle ingots weighing more than 300 tonnes. This process—called *cogging*—produces a *bloom*—an elongated billet of steel in which the cast structure has been broken up, the cavities filled and the grain induced to flow along its length.

The workpiece is held in a manipulator, which rotates and feeds it between the tools of the press—hence this type of forging is called *open-tool forging*.

To elongate the shaft further, it is reheated and then swaged or pressed between two arch-shaped tools which, when closed together, form a hollow cylinder. The tools may have two or more arches of different sizes so that the workpiece can be moved from the larger to the smaller as its diameter reduces. When the workpiece nears its finished dimensions it is *heat-treated*—heated and cooled repeatedly—to develop the optimum metallurgical properties.

Open-tool forging can be used to produce other shapes, such as a slew ring on a large crane or excavator. If the product is to be round, such as a sphere for a ball valve in a gas pipeline, the ingot can be *upset*—or have its diameter increased instead of being cogged into a long shape.

Open-tool forging brings the workpiece to only an approximate shape of the finished component, so much time and materials are wasted in machining. A more accurate technique—*closed-die* forging—employs a mould or closed dies into which the hot metal is squeezed. The dies are the top and bottom halves of the forging tools, and when brought together they form an accurate mould.

Die forging is more easily controlled on presses but it is still used extensively on hammers, which have been refined with the aid of modern engineering. In hammer forging, the top half of the die is attached to the hammer—called the *tup*—and the bottom half to the anvil. The tup, which might weigh as much as 18 tonnes, is raised on guides and the slug of hot metal to be forged is placed in the bottom half of the die. When the tup is released, it falls under gravity and strikes the metal, causing it to flow into the shape of the die. After several blows, the metal fills completely the impression in both halves.

Some large gravity-drop hammers are power assisted, using steam or compressed air not only to raise the tup but also to force

Machine Technology: Metallurgy

KEY *(below)*
1 Ingot on station
2 Ingot is deformed and descaled
3 Billet is placed in blocking pot
4 Diameter increased
5 Punch positioned
6 Blocker is punched
7 At trim station
8 Blocker is trimmed
9 Ready for extrusion

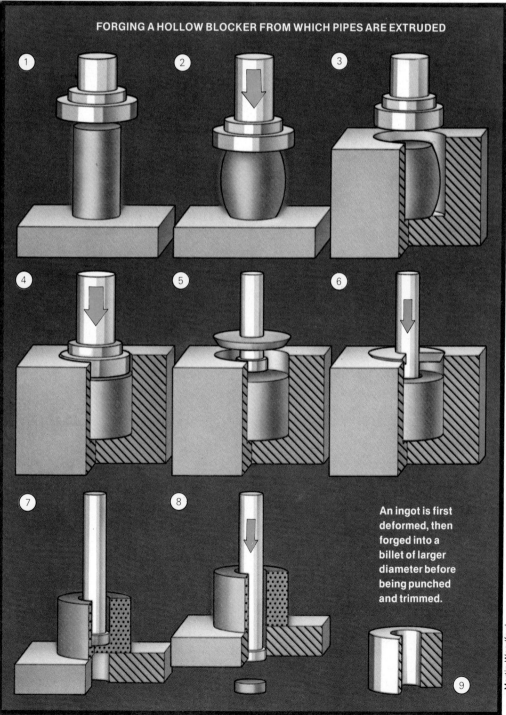

FORGING A HOLLOW BLOCKER FROM WHICH PIPES ARE EXTRUDED

An ingot is first deformed, then forged into a billet of larger diameter before being punched and trimmed.

it downwards. These hammers have been in use for decades for open-tool work but for closed-die forging—also called *drop stamping*—hydropneumatic systems are more popular. They employ hydraulic power to raise the tup and at the same time compress a volume of air. When the tup is released the air blasts it downwards forcefully, so fewer drops are required for a forging.

It might not be possible to coax the metal to change shape from a cylindrical billet into a complicated shape in one die, so intermediate stages are necessary. In this case the same die block incorporates several moulds, each one shaping the metal a little closer to the finished product; between hammer blows the slug is moved progressively from one die to the next. If the size of the forging is not too large, a single hammer blow might be sufficient to form a first forging or *preform* in the first die while the finished component is being made in the second.

The metal slug is never the right size to fill the die, so the die is made with a *flash land*—a relieved area outside the mould proper—to accommodate any excess. When the forging comes out of the last die, it has a flash of metal where the join in the die was. To remove this the forging is placed into a clipping press in which the bottom die is an outline of the true shape of the forging. The tup carries a die which punches the forging through the bottom die, leaving the flash.

Motor components

Drop stamping is commonly used to manufacture motorcar components, such as gear blanks, steering arms and idler wheels. As many as seven intermediate stages might be carried out on different machines: gear clusters, for example, have gear wheels of different sizes on a single shaft; they can be made in up to six sizes on a horizontal upsetter, some of which can produce forgings at a rate of 120 per hour.

The upsetter has a stationary and a moving gripper die, each with six impressions. In operation, the grippers are opened and a hot bar is inserted into the first impression with a predetermined amount projecting through into the machine. The grippers close, holding the bar, while a ram pushes against the protruding end and upsets the bar into the die impression. The forging is then rolled into the next impression and the bar diameter is again upset to form gear wheels of progressively larger diameter.

Some components are best produced on a production line. The manufacture of crankshafts, for example, begins with twin

Machine Technology: Metallurgy

induction heaters which heat the billets to forging temperature within a few seconds. The billets move on by conveyor to a reducer roll, where preforms are made, and then via another conveyor to an 8,000-tonne press. The crankshafts are forged in two or three operations, depending on the size and complexity of the design.

From the large press, the forgings go on another conveyor to be clipped on a 750-tonne press. Next they go by conveyor to be *coined*—a finishing process which smooths out the rough edges where the flash has been trimmed off and sizes the crankshaft as accurately as possible. Finally the forgings cool on another conveyor.

The prime goal

The production-line arrangement above would be termed semi-automatic because it needs men—both to load the forgings into the presses and to operate the controls. It is an extremely arduous job carried out in hot, dirty and noisy surroundings—not the conditions to attract skilled men operating costly machines. Consequently, full automation is the prime goal of forging-tool designers.

Near Turin, in Italy, a completely unmanned forging line at Forge of Fiat is operated by computer-controlled robots. In a typical layout, a conveyor moves a bin of steel billets automatically to a position where a robot can pick them out and load them on to the conveyor leading to the induction heater. Another robot then feeds the heated billets on to the die bed of the forging press. During forging, the billets are supported in the press by two robots, who grasp the component at each end while lifting it from die to die. An unloading robot grips the completed forging and transfers it to the clipping press.

New materials are always being developed for the vast range of modern industries—from aviation to nuclear power—which use forgings. In a gas turbine, such as a jet engine, some components must be light but strong, and must also retain shape in the intense heat. Alloys—which comprise two or more metals or a metal and other elements—have been developed expressly for high strength in such conditions, and their properties are enhanced by forging.

The first row of blades in the rotor of a gas turbine operates at temperatures ranging from 600°C to above 1,000°C at high rotational speeds. The energy stored in the high pressure (HP) turbine assembly of, for example, a Rolls Royce Conway engine at take-off is equivalent to that of a 25-tonne vehicle travelling at 53 km/h.

Nickel-based alloys, which are generally used for turbine blading, became steadily more complicated as the working temperatures of engines increased. But although the new alloys performed better at higher temperatures, they became increasingly more difficult to work. Today, even the most temperature and stress-resistant turbine blades are cast and forged.

To enable turbine blades to work in the searing gases of a jet engine, they are equipped with internal cooling passages through which air passes. This, coupled with the fact that they have to be made from some of the toughest alloys available, poses formidable manufacturing problems. One method is to forge each blade deliberately out of shape, drill the cooling holes, and then coin the blades to the correct shape.

As the surface area of a forging becomes larger, the forces needed to forge it increase astronomically. One critical jet-engine component with a large surface area is the disc to which the turbine blades are attached. In France, Interforge produces these components on a mammoth 65,000-tonne single-ram press. One of the advantages of forging with a press rather than a hammer is a significant reduction in the strain rate—the speed at which the metal moves in the dies—and this in turn helps to enhance important metallurgical properties.

Strain rate can be reduced even further by *isothermal forging*, in which the dies are at the same temperature as the workpiece and low-tonnage presses can be used. The forging temperatures are in the range of 960–1,050°C for titanium and 1,050–1,150°C for nickel-based alloys. Very strong molybdenum alloy dies are used at these high temperatures but even so the work must be carried out in a vacuum or an inert atmosphere to prevent the dies wasting away in the oxygen of the air.

In the USA, Pratt and Whitney has made use of the *superplastic* property of nickel-based alloys from which turbine discs are made in the Gatorizing process, which gives a strain rate many thousand times slower.

The Gatorizing process employs superalloy powder, which is extruded under controlled conditions to make it superplastic. The extrusion is then forged to shape isothermally in presses that can control the strain rate accurately but need to have capacities in the 1,650–8,000 tonne range.

The Gatorizing process also introduced the revolutionary idea of starting the forging with a powder and not a solid lump of metal. It was then a relatively simple step to develop

a forging technique which formed the powder closely to the finished shape and compacted it isothermally.

In one method used by GKN to produce connecting rods for the Porsche 928 sports car, the first step is to compact an accurate preform out of the metal powder in a die press. This results in fragile 'green' compact which can be handled only with care. The compact is placed in a furnace and is toughened by the heat.

Automated forging, monitored remotely *(far left)*, relieves workers of the heat and noise of the forging hall. Some forgings—such as turbine shafts *(below left)*—are so massive that they can be manipulated only remotely. But even smaller ones—such as valve bodies *(left)* and turbine blades *(below)*—are best automated.

Often sintered (fired—like pottery) components are used for non-critical components, but for heavy-duty applications they are heated to forging temperature. To reduce the cost of dies, the powder is placed into a flexible mould—usually a tough rubber—which is sealed and loaded not into a conventional press but into a chamber of pressurized liquid. Because the mould is flexible, the pressure acts on the powder from all directions. For this reason, the technique is called *isostatic pressing*. After pressing, the green compacts are sintered in a vacuum, then forged even in between open tools.

Isostatic pressing is well suited to making small quantities of large, expensive components. Turbine discs are just such components, and engineers have devised a production route in which the disc is pressed to near the finished shape. The difference is that the pressing is done hot, the process being known as hot isostatic pressing or HIPing.

Gas pressurization

The mould in this case is steel or glass and, after de-gassing, it is sealed—usually by welding. Instead of a liquid, the mould is pressurized by a gas (usually argon) in a heated press. The Quintus press used in Sweden by ASEA consists of a steel cylinder, open at both ends, wound with thousands of turns of pre-stressed wire.

After the mould and the heaters have been loaded into the cylinder, it is loosely plugged top and bottom and the whole assembly slipped into a massive oblong ring. This ring, also wound with pre-stressed wire, stops the top and bottom plugs blowing out.

HIPing can be used to produce a billet, which is then forged or Gatorized, or to produce a preform, which again is forged. It can also produce a near-finished shape, in which case it is called direct HIP or asHIP.

Direct HIP is potentially the most attractive process because it eliminates the forging stage altogether. Aero engines were running with asHIP discs in the early 1980s but engineers accepted that their properties might be inferior to forged discs. Nevertheless, techniques are constantly advancing so this is unlikely to remain true for long.

Forging has come a long way from the blacksmith's anvil, due to the need to produce stronger and more competitive components for industry. Increasingly, computers are helping to improve design, production and metallurgical control. But the heart of the process is still the same as it was in the Iron Age: shaping hot metal to fill the needs of Man the technologist.

Frontiers: Palaeontology

Man's origins: the biochemists' story

Left According to the fossil record, none of Man's ancestors seem to have been able to brachiate—swing from tree to tree—like this orang-utan. Biochemists think they may have solved the enigma, in the process disputing palaeontological evidence *(below)*.

As far as Man's ancestors are concerned, the fossil record stops abruptly four million years ago. We have fossils of upright-walking hominids from Africa (including the famous 'Lucy' skeleton) that are 3.75 million years old, but before that there is nothing that can certainly be called human. Such a dearth of palaeontological evidence for Man's ancestry has led another branch of science—biochemistry—to join in the hunt for clues. The findings in this field so far are startling—and contradictory.

Many palaeontologists have singled out one creature—*Ramapithecus*—as an early human ancestor on the basis of certain detailed features of its teeth and jaws. *Ramapithecus* is one member of a group of ape-like animals called *ramapithecines* which inhabited Africa and Asia between 8 and 14 million years ago. Central to the argument that it is an early hominid is the fact that it may have walked on two legs to gather food more easily.

Acceptance of the argument, however, entails an evolutionary scheme in which our ancestors went straight from quadrupedalism to bipedalism, without ever developing a form of locomotion known as *brachiation*—swinging from branch to branch.

Some palaeontologists find this difficult to swallow; they feel that Man and his immediate ancestors—the apes—must share common brachiating ancestors which the fossil record has yet to show up.

Brachiating ancestors

Brachiation is a special feature of the apes, most highly developed in the gibbon and the orang-utan, but clearly evident in our nearer ancestors, the chimpanzee and the gorilla. Human beings, too, have some features that resemble those of brachiators: for example our shoulder joints are very flexible, and our arms can take our full weight.

It must be said that not all shared characteristics are necessarily the result of a shared past; hominids could have learned to swing from tree branches after their ancestors diverged from those of the apes. Nevertheless, the idea of a common brachiating ancestor remains appealing since it is simple and fits all the known facts apart from *Ramapithecus* (which, if it were walking upright 8–14 million years ago, must be assigned to the status of an evolutionary 'dead end' by this scheme).

Conventional wisdom based on the fossil record (and disregarding the ambiguous evidence of the ramapithecines) places the split of Man and ape anywhere between 20 million years ago, when the quadrupedal ancestors of the ramapithecines lived, and 3.75 million years ago when 'Lucy' appears. Palaeo-anthropologists generally favour an early date, about 15 million years ago, which allows *Ramapithecus* as an early hominid, and gives Man a long period of development after separation from the apes.

Most of the specialists in this field have a deeply held belief that this long period of independent evolution must have taken place for Man to develop his unique features. For the same reason they tend to reject what evidence there is in favour of a brachiating human ancestor, since this would imply that the link with the apes is more recent. But now, new evidence produced in the field of biochemistry suggests that just such a link existed despite the lack of fossils.

Below left Examining the protein sequences of Man's relatives (like the knuckle-walking chimp *below*) is just one of the areas of research throwing new light on the time we 'split' from our ape predecessors.

In the past 25 years there have been tremendous advances in biochemical techniques which have allowed detailed studies of the chemical constituents of the body. Most important of these from the point of view of Man's origins are *proteins*, which are vital structurally and catalyze all the chemical reactions of the body as enzymes.

Protein analysis

The first stage in protein analysis is to determine if the protein consists of more than one *polypeptide chain* (a chain of amino-acid residues); if it does, it must be treated chemically to split the chains, which then have to be separated. Once a solution containing only one type of polypeptide has been obtained, the overall amino acid content is determined using an *automatic amino acid analyzer*. Then the chain is treated with enzymes that are known to break the polypeptide chains at particular points. The short polypeptide lengths obtained are separated by chromatography, and their amino acid content again determined. This process may then be repeated using another enzyme to break the chain in different places.

The next step is a process called *end group analysis*, which relies on the fact that one end (terminal) of the polypeptide chain carries a carboxyl (C) group ($-CO_2-$) while the

Frontiers: Palaeontology

other carries an amino (N) group ($-NH_3+$). A variety of different enzymes or chemical reagents can be used which react either with the N-terminal or the C-terminal amino acid residue, splitting it off from the chain; the amino acids can be identified separately by chromatography as they split.

Although in a few cases end group analysis can give a clear picture of the amino acid sequence, the amino acids tend to split off at different rates. This makes it necessary to combine the above processes with other, more complicated, experimental steps—but still using variations of the basic methods—until enough 'clues' have been obtained to piece together the amino acid sequence and thus discover the protein's unique identity.

With sequences for a variety of proteins from many different animal species now established, it has been possible to see how protein sequences have changed during the evolutionary process. For example, the sequence of the protein cytochrome c has been obtained for a large number of animals and an evolutionary tree can be constructed on the basis of the relationship between the sequences, placing those that differ by only one amino acid residue closest together, those that differ by two amino acid residues a little farther apart, and so on.

Computers can be used to analyze the data and construct an evolutionary tree. In general the tree that emerges agrees fairly well with the conventional one, based on palaeontology and comparative anatomy.

The reconstruction of a tree makes it possible to work out the ancestral sequence—that is, the most likely sequence of amino acids in a particular protein in the common ancestor of a group of animals.

Right and *far right* The fact that humans and apes share many brachiating characteristics—i.e. flexible shoulder joints—suggests they may have shared a brachiating ancestor.

One of the most exciting discoveries in the field of genetic research is that the nucleotide bases that make up the DNA chain—the genetic code of an organism—undergo spontaneous changes (mutations) that seem to proceed at a regular rate. The significance of this observation is that the sequence of nucleotide bases codes directly for the sequence of amino acids in protein. If differences between proteins occur at a constant rate it should be possible—by calculating the difference between the proteins of two species—to calculate how long ago they diverged from a common ancestor.

This 'molecular clock' theory is based on comparing the proteins of many different pairs of animals whose times of divergence are known from the fossil record and plotting the results as a graph to deduce the rate of nucleotide substitution. If the number of nucleotide substitutions affecting the same set of proteins is then determined for another pair of species, their time of divergence from each other can be read off the graph.

When it comes to comparing the human and ape proteins a major problem is finding reference points on which to plot a graph in the first place, since there are few reliable dates for divergence times in primate evolution which can be used for comparison.

Clearly, the results include a large margin of error, but even taking this into account the disagreement between the molecular clock figures and the conventional date is startling: the biochemists give a mere five

Below Each of the protein molecule's polypeptide chains contains about 150 amino acids. Before these can be analyzed the polypeptide chains must be separated by enzyme treatment and isolated (5). Each chain (6) now goes through 3 different processes.

THE ROAD TO PROTEIN ANALYSIS

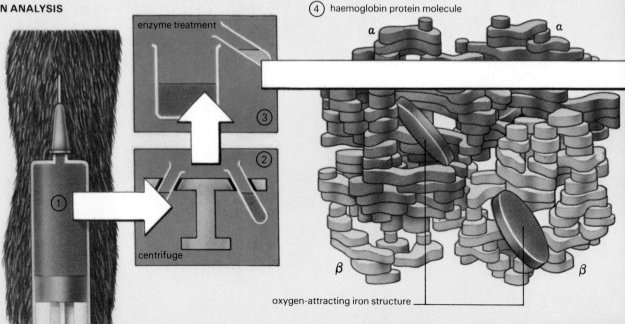

Right The laborious process of protein analysis starts (1) with a blood sample—in this case taken from a chimp. The blood is then centrifuged (2) and treated (3) to extract a single substance—in this case haemoglobin. Closer examination of a single haemoglobin protein (4) reveals 4 polypeptide chains—2 alpha and 2 beta.

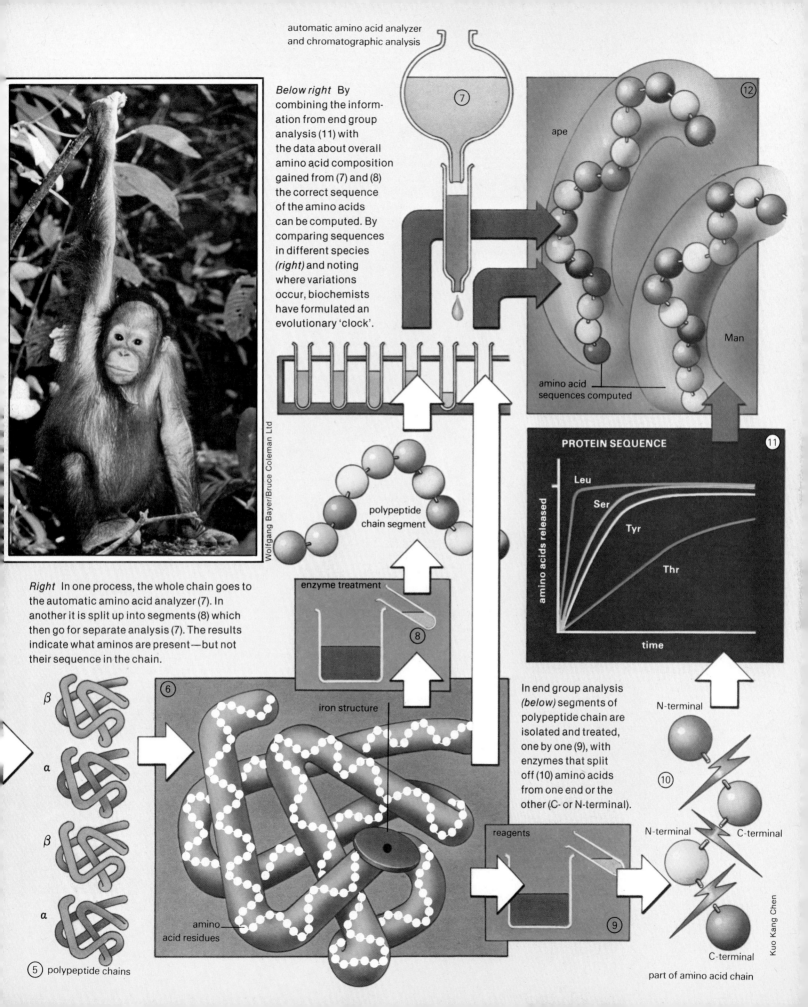

Frontiers: Palaeontology

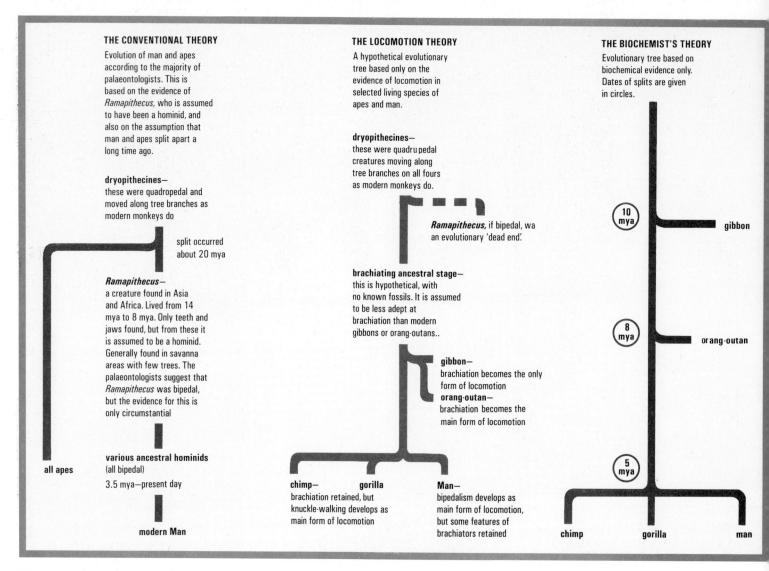

million years as the time since the ancestors of the African apes (gorillas and chimpanzees) and those of humans separated—substantially less than the 15 million which most palaeontologists favour. Their results suggest that we are as closely related to gorillas as to chimps, less closely to the orang-utan from which we split off eight million years ago, and less again to the gibbon who last shared a common ancestor with Man about ten million years ago.

To substantiate these figures, the biochemists have investigated the biochemical relationships between humans and apes by other methods that are less time-consuming than protein sequencing and can therefore be applied more extensively. One important technique is based on the immunological reaction in which an animal produces an *antibody* to a foreign protein (the antigen) as part of the body's defence mechanism against infection.

Using an extremely complicated procedure known as *micro-complement fixation* it is possible to compare how antibodies react to a certain antigen in one animal (say, a chimpanzee) with how they react to that same antigen in another animal (say, a human).

This resulting figure is known as the *index of dissimilarity* and the larger it is, the farther apart the animals concerned are in evolutionary terms. It has been shown that the index of dissimilarity between chimps and humans is directly proportional to the amino acid residue differences that are revealed by protein sequencing.

The results of another method for distinguishing proteins, known as *electrophoresis*, also show a direct relationship to sequence differences. The method separates proteins on the basis of their electrical charges, which reflect their amino acid composition. Electrophoresis has the advantage of being able to distinguish a usefully large number of proteins at the same time.

The latest biochemical analysis technique for investigating Man's ancestry deals directly with the genetic material—DNA. It does this by measuring the degree of change in the four nucleotide bases (guanine, cytosine, adenine and thymine) which go towards making up the DNA molecule.

DNA analyzed

DNA is composed of two strands which are held together by hydrogen bonds between the bases. The bases have a strict pairing system in that guanine always pairs with cytosine on the opposite chain, and adenine with thymine. Thus if a single strand of DNA from one animal is artificially paired with one from another animal not all their bases will match up because mutations will have changed some of them. Consequently, there will be fewer hydrogen bonds and the pairing will not be as strong as it is normally.

Left A comparison of palaeontological and biochemical timescales for Man's ancestry with a hypothetical family tree based purely on locomotive features. *Right* Although the fossil hunters have yet to confirm it, we may be more closely related to apes *(below)* than once thought.

What is actually observed is that hybrid DNA separates into individual chains more readily than normal DNA when the mixture is heated. Trials with synthetic DNAs of known sequences have shown that for every one per cent change in the bases, the temperature at which the chains dissociate drops by 1°C.

When human and chimpanzee DNAs are compared in this way the difference between them is very small—about one per cent. The technique, known as DNA *annealing,* has so far been applied mainly to primates and rodents, both of which lack secure fossil dates for divergence times, so it is difficult to translate this figure into a time value. However, the results so far point to a divergence at less than ten million years ago.

The major limitation of the 'molecular clock' in timing Man's ancestry is not the biochemical data itself, but the unreliability of the dates obtained from fossils against which the clock is calibrated. Although the fossilized remains can generally be dated accurately, the question of where they belong in the evolutionary tree is open to a variety of interpretations—and many different times of divergence might be given on the basis of exactly the same evidence. A great deal of confusion has arisen because the biochemists involved have often not realized just how unreliable the dates given them by palaeontologists are. Likewise, the palaeontologists, impressed by the accuracy of the biochemists' techniques, have often forgotten that the precise figures their experiments produce conceal the original, and very dubious, fossil-based times of divergence.

As a result of the confused situation, it is by no means absolutely certain that the rate of nucleotide substitutions is constant —although most of the evidence points to it.

Some workers have suggested that the substitution rate depends not on absolute time, but on the time between generations of the animals involved; this would reconcile the dates obtained for the human-chimp split with the conventional divergence time of 15 million years, since the generation times of the apes—and of Man especially—are much longer than those of other primates.

No differences

A few experiments do seem to show that generation time is what counts. But the main protagonists of the 'absolute time' relationship, Allan Wilson and Vincent Sarich of the University of California, have carried out a series of tests designed to measure the relative rate of change in two lines of different generation time, and they have been unable to show up any differences.

Whatever the causes for our differences from the gorilla and chimpanzee, the 'molecular clock' suggests that we separated from these ape cousins only five million years ago, and even allowing for all the possible errors in interpreting the experiments, a time of more than ten million years is highly improbable. This rules out a bipedal *Ramapithecus* as an ancestor at 8–14 million years ago, and replaces him with one that looked far more like the modern African apes—presumably a forest-dwelling species that swung from tree to tree.

Clearly the biochemical evidence and the tenuous anatomical evidence for a tree-dwelling human ancestor fit together and point to a similar evolutionary scheme. And as forest-dwelling species get into the fossil record far less often than others, because the right conditions for fossilization are rare in forests, this might explain why our early ancestors have proved so elusive.

Military Technology: Nuclear Weapons

The people killer

In an age of nuclear weapons and massive conventional firepower the consequences of a major war are almost unimaginable. A nuclear warhead of the type deployed by most of the major powers would cause death and destruction on a huge scale, killing indiscriminately and making whole areas of the world uninhabitable. So now a new generation of warheads and delivery systems—popularly known as the neutron bomb—is being developed to concentrate the destructive effects of an atomic blast in a small area with a reduced risk to non-combatants. Unfortunately, although the theory may be cosy, the realities of neutron bomb warfare are just as horrifying as those of conventional nuclear conflict.

The neutron bomb is becoming accepted as part of a NATO defence strategy that has evolved since the early 1960s, and central to this strategy are three facts. The first is that NATO are outnumbered by the Warsaw Pact, both conventionally and in terms of nuclear capability. The second is that NATO could not hope to contain a steamroller-like Soviet assault using conventional weapons alone. And the third is that a war of this type would have to be fought on NATO territory, with civilians in the battle area and the certainty of heavy casualties and unacceptable levels of damage in towns and cities.

In the event of an invasion of Europe NATO's plans involve the use of conventional forces to fight a delaying action and channel the invaders into boxed 'killing zones' where they can be destroyed en masse using nuclear weapons. The problem for NATO commanders and planners is that the type of warhead most commonly deployed in Europe up to now—on both sides—is an indiscriminate killer which will cause huge amounts of 'collateral' damage around the

Above Delivery systems like the Lance missile or Tornado *(far right)* make the use of small nuclear weapons highly attractive to commanders. But millions of people *(top)* are afraid that they also make the use of strategic weapons far more likely in the event of a major European conflict.

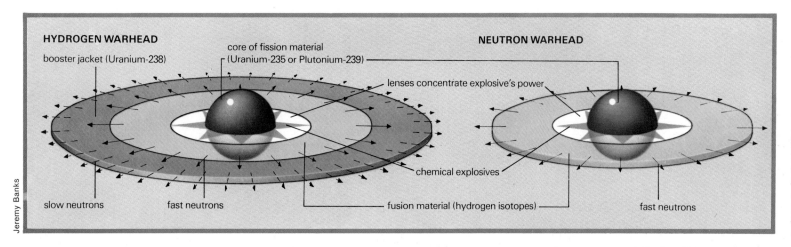

immediate target area of the warhead.

What they feel they need, and what the neutron bomb seems to supply, is a weapon with concentrated killing power which can be accurately placed on the battlefield to destroy the enemy without needlessly devasting surrounding towns and villages.

The Soviets, for their part, feel threatened by the NATO presence in Western Europe and have expanded their forces accordingly. Their prime concern is self-defence and the protection of Eastern Europe, a task they approach with the attitude that 'attack is the best form of defence'.

Integrated arsenals

Although they have experimented with a similar type of weapon, the Soviets have no real need for the neutron bomb. They do not believe that they will have to fight on their own soil, and so do not need the comparative delicacy offered by this new weapon. As a result they do not need to distinguish between conventional and nuclear warfare—with ultimate victory their only goal in any major conflict, they have integrated their nuclear and conventional arsenals right across the board.

A European war could be fought using conventional weapons alone, but only if NATO holds its own against invading forces. Many people fear that once NATO succumbs to the temptation to start using tactical (that is, comparatively small) nuclear weapons like the neutron bomb, then the possibility of the conflict escalating to an all-out strategic bombardment with catastrophic results becomes far greater.

Quite apart from the wholesale slaughter that would follow any kind of nuclear exchange, many are haunted by the spectre of a sizeable portion of Western Europe left totally uninhabitable for generations to come. Even if the nuclear exchange were

Above In a nuclear warhead, chemical explosives set off a fission reaction in the core, which in turn sets off a fusion reaction in the surrounding material. This releases huge numbers of fast-moving neutrons, which set off a fission reaction in a uranium jacket. The neutron bomb has no such jacket, so its radiation levels are considerably higher.

limited to battlefield weapons alone, huge tracts of West Germany would be devastated and much of that country would cease to exist as an economic and social entity.

Nevertheless both sides continue to develop their nuclear and conventional capabilities, the Soviets making up in numbers what they lack in technological prowess and NATO relying on high efficiency and advanced weaponry. To complement its new generation of weapons NATO is also introducing delivery systems that can hit the enemy with almost surgical precision, most of them electronically guided missiles of one kind or another.

The most prominent type of theatre missile currently employed by NATO is the Pershing I, of which 180 are deployed in Central Europe. The Pershing has a maximum range of 720 km and carries a 400 kT warhead—equivalent to 400,000 tonnes of TNT, the most widely used explosive in modern weapons.

A standard nuclear charge of this size will create a fireball 1.5 km across and a thermal sphere up to 9 km in diameter capable of melting metal. Across a 16 km area wood will spontaneously ignite and across 25 km people will suffer third-degree burns. Inside this area, all life will be extinguished. In an area 55 km across first-degree burns will be inflicted and even outside this zone injury will be widespread. Multiplying these effects

Military Technology: Nuclear Weapons

Left The pin-point accuracy of the cruise missile makes the use of neutron warheads far more likely than before.

several times over highlights the dilemma facing NATO commanders, who would have to order such destruction to halt enemy advances into their own territory.

All nuclear weapons have three effects: heat, blast and radiation. In an ordinary thermo-nuclear bomb, 50 per cent of the energy is released as blast and only 15 per cent as radiation.

However, there is one type of nuclear weapon which releases 80 per cent of its energy in the form of high-energy neutrons—the neutron bomb.

As well as neutrons, radiation from an ordinary nuclear weapon includes beta particles which travel great distances, plus short-lived alpha particles, gamma rays and protons which die quicky. Because they have a short range, neutrons transfer more energy to the target area than do the other forms of radiation so they are more lethal. Neutron bombs increase their lethality by propagating neutrons which are more than ten times as dangerous as beta particles or gamma rays.

Moreover, neutrons can travel straight through protective armour and actually 'spill' particles from the atomic structure of metal, spraying them around the interior of a tank to be absorbed in the bodies of the personnel inside. In this way, the blast damage can be significantly reduced but the radiation selectively increased to an intensity where it kills outside the range of the comparatively small shock wave and heat pulse.

Small but effective

Whereas weapons with explosive yields of 10 to 500 kT are standard stock, neutron weapons of only 1 kT yield are necessary for battlefield use. A neutron weapon of this size will produce 18,000 rads (a unit of radiation) up to 400 m away, incapacitating within five minutes everyone inside that radius, and causing death in a matter of hours. At a range of 650 m, people would receive 8,000 rads and die within two days. At 800 m doses of up to 3,000 rads would be received, causing vomiting, a falling red blood cell

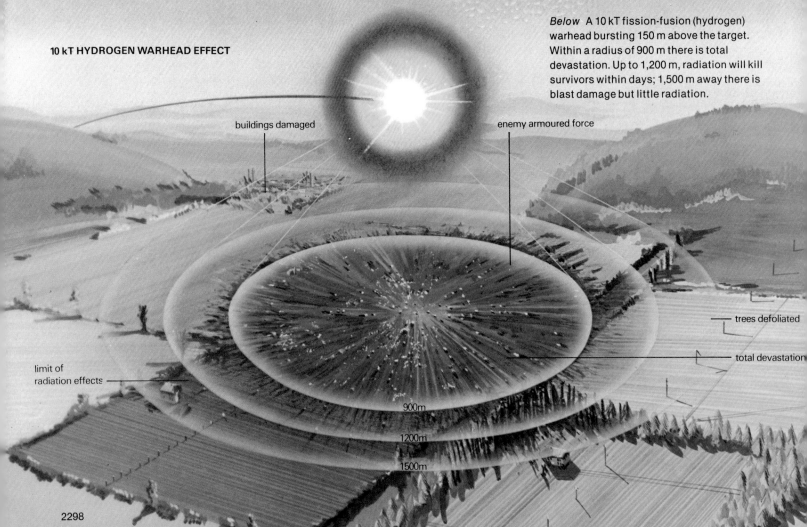

Below A 10 kT fission-fusion (hydrogen) warhead bursting 150 m above the target. Within a radius of 900 m there is total devastation. Up to 1,200 m, radiation will kill survivors within days; 1,500 m away there is blast damage but little radiation.

10 kT HYDROGEN WARHEAD EFFECT

count, internal bleeding, collapse of the nervous system, and organic failure followed by massive heart attack. Death would ensue up to five days after the attack.

At nearly 1,000 m, radiation doses would average 650 rads, causing partial collapse of the nervous system. Casualties would need immediate therapy involving blood transfusion, special diet, antibiotics, and a bone marrow transplant. Even if all these services were available immediately, 70 to 95 per cent of them would still die several weeks later.

Other effects down to a dose as low as 150 rads, will completely sterilize the patient. Doses of 50–150 rads picked up by people on the edge of the battlefield would result in genetic damage and possible mutation. And recipients of 100 to 200 rads would have only a 60 per cent chance of surviving more than two months.

Because the debilitating effects of neutron bombs kill people more slowly, mass panic and confusion among the enemy are considered to be a side benefit of this type of weapon. All but a very few within the radius of the fireball will take at least one day to die and the problems involved in clearing these people from the battlefield would assume monstrous proportions—perhaps greatly influencing the final outcome of the conflict.

Some think that the terror of neutron bomb warfare could invoke 'suicide' missions where soldiers hurl themselves upon the enemy without recourse to normal logic and discipline, preferring to die quickly in a fire-fight than slowly and in excruciating agony. Either way, it would completely transform the battle.

Neutron warheads could be delivered by air or missile but the preferred method would be to use artillery. Pershing missiles or the 110 km (70 miles) range Lance might be used to carry neutron heads but the US Army would like to fit them to the very accurate M2A1E1 self-propelled howitzer which has a range of 16.8 km (10 miles). This would be suitable for a warhead shaped to deliver lethal radiation over an area tucked

Right The neutron bomb renders most NBC (nuclear, biological, chemical) defence equipment obsolete overnight.

Below A 1 kT neutron warhead bursting 150 m above the target. A 400 m radius is devastated and radiation will kill tank crews 750 m away immediately. Within 900 m victims die in six days, and within a 1,200 m radius they die in a matter of a few weeks.

Left Both the superpowers are manoeuvring all the time, trying to gain the advantage. The nuclear cannon *(below left)* and Pershing II missile *(below)* allow NATO plenty of firepower, matching the Soviet numerical superiority.

in between two closely spaced towns.

If deployed in this way, neutron heads would probably be available to a standard 12-gun battalion should the density of opposing armour make it attractive to use them. However, because very few people would die immediately, several thousand tank crews who knew they were doomed to die in a horrifying manner might unleash a fanatical and unstoppable assault.

It is quite possible that the widespread havoc and devastation would reverse any advantage initially sought by invoking the use of neutron heads. It might actually cause more civilian deaths than the use of neutron heads would theoretically prevent.

Of perhaps more immediate concern to potential enemies, however, are the new theatre systems NATO is committed to deploy—unless arms control negotiations achieve agreement to reduce or reverse the build-up of these weapons.

NATO plans involve the removal of 1,000 old nuclear weapons from Europe and their replacement with 572 new delivery systems: Pershing II and the ground-launched cruise missile, or GLCM. The new Pershing can send nuclear charges across a maximum range of 1,600 km compared with 720 km for Pershing I, the missile it will replace on a one-for-one basis. Moreover, the latest models carry an area-correlation radar guidance system. This requires the missile's two propulsive stages to fire the warhead into the general area of the target, making it at that point about as accurate as Pershing I.

From a height of approximately 5 km, however, a small radar unit in the nose of the warhead scans the surface below and compares the information with a stored 'mosaic' of the precise area into which it should descend. The warhead has control surfaces which respond to information that moves the device back on track from the marginal errors which are expected to have built up during the flight.

From a height of 1 km the warhead is on its own and achieves a free-fall accuracy of about 30 m. It is precisely because of this accuracy that the yield of the warhead carried by Pershing II has been reduced from a selectable range on Pershing I of between 60 and 400 kT to a range of between 1 and 50 kT. This permits Pershing II to be used close to

Military Technology: Nuclear Weapons

friendly troops, the high accuracy and low yield producing little collateral damage.

An alternative warhead would be a very low yield 'earthquake' device designed to penetrate the ground to a depth of 50 m and then explode, spewing highly radioactive dust and debris across a wide area. In another application Pershing II could, because of its extreme accuracy, throw a clutch of 76 high explosive charges on to an airfield, tearing up the runway and destroying buildings.

'Bang per buck'

The main attraction of missiles like Pershing II is their comparatively low cost, giving 'more bang per buck' than conventional aircraft dropping free-fall nuclear bombs. A modern strike aircraft can cost $30 million while Pershing II costs less than $1 million. This also applies to the GLCM variant that NATO plans to deploy in Europe.

At less than $1 million each, the GLCM is seen as an attractive way of both restoring the balance in airborne delivery systems and eliminating human losses in manned aircraft over hostile territory. The cruise missile is actually a pilotless flying bomb. Travelling at about 1,000 km/h with a maximum range of 2,500 km, it can fly very low because it uses radar that 'reads' the ground and controls the flight path and altitude. It is extremely accurate because it scans selected sections of the terrain across which it flies to compare its flight path with the course programmed in its computer and then performs the necessary corrections as it goes.

The GLCM will be deployed in transporter-erector-launcher (TEL) vehicles, each of which can carry and fire four cruise missiles. Flung into the air by a rocket motor which falls away seconds later, the device flies like an ordinary aeroplane on the power of a small turbojet set in the rear fuselage. Its warhead is armed only when it leaves friendly airspace, ensuring that if the missile crashes or is shot down by an intruding aircraft it will not detonate the 160 kT charge.

NATO says the Pershing II and GLCM are essential to counter what it sees as a massive arms build-up in Russia and the Warsaw Pact. Russia says that the new cruise and Pershing II systems represent a new threat which comes in addition to Britain's expansion of its own nuclear forces with US Trident missiles.

Neither side may have a totally logical argument but each believes it faces a perilous future without new and improved weapons. There are about 14,000 nuclear warheads, depth charges and land mines in Europe, but the only real value of a figure like that is to instill a feeling that people, and not weapons, will make that troubled area a safer place to live in future.

The late Sir Winston Churchill is quoted as saying, 'Jaw, jaw is better than war, war'. And since the 1970s America and Russia have tried, through the SALT (strategic arms limitation talks) treaties, to defuse a potentially explosive situation. But neither party has really made a serious reduction in the forces deployed throughout Europe. The START (strategic arms reduction) talks which began in 1981 are the latest attempt to achieve this for the sake of mankind.

Below 'Jaw, jaw is better than war, war'—especially nuclear war. The START talks between the USA and the USSR which began in 1981 are aimed at reducing the number of weapons deployed by each side in Europe. As long as people can meet around a conference table, it is hoped that Europe will be safe from the horrors of a nuclear war.

Electronics In Action: Musical Instruments

Wired for sound

Denis O'Regan

The conventional acoustic guitar has been around for at least 600 years, but the technological research that surrounds its much younger electric cousin has caused rapid and dramatic change. As long as scientists and musicians can work together harmoniously, the electric guitar will benefit from the rare combination of traditional craftsmanship and technological innovation.

In 1931 Adolph Rickenbacker's Electro String Company in California produced the world's first electric guitar, designed by George Beauchamp and Paul Barth. It was a lap-steel guitar, now known as the 'frying pan' because of its long neck and round body made of aluminium and, later, Bakelite.

Four years later the Gibson Guitar Company of Michigan produced its first electric semi-acoustic guitar, the ES150, and in 1948 Leo Fender, a radio repairman in California, made the first commercial solid-bodied electric guitar, the Broadcaster (later renamed the Telecaster). Guitarist and inventor Les Paul collaborated with Gibson to introduce the famous Gibson Les Paul in 1952, and two years later Fender launched his stylish Stratocaster, which has since become virtually an 'industry standard' rock 'n' roll guitar.

Since this burst of creative energy from designers, technicians and musicians, considerable variations and improvements on these basic themes have appeared, although some changes made in the name of innovation have been mere gimmickry. Setting these aside, there are two areas of real development that concern most electric guitar producers today: the guitar synthesizer and the search for new materials.

Straightforward instrument

The electric guitar is essentially a very straightforward device. Vibrating strings struck by the player are transformed into electrical energy by a pickup or pickups, and this energy is in turn amplified by an external amplifier and loudspeaker. A wooden body and neck, either in one piece or else secured to each other, support the strings.

The sounding length of the strings is stopped at the body end of the bridge (offering varying degrees of control over horizontal and vertical positioning of the strings), and at the headstock end by the nut. Slots cut into the headstock direct the strings accurately

Above Adolph Rickenbacker displays the first electric guitar, affectionately known as the 'frying pan' due to its shape. *Opposite* Fee Waybill of The Tubes with his custom-made guitar.

The 1960 Gibson Les Paul Standard *(above left)* and the 1958 Gibson J200 acoustic guitar *(above right)*. *Left* The 1948 Fender Broadcaster which was later renamed the Telecaster.

to the machine heads, peg and gear mechanisms which enable the player to tune the strings. The pickup(s) and associated controls are located on the body.

The pickup is a transducer—a device which converts one form of energy (vibrating strings) into another form of energy (an electrical signal). The metal strings vibrating in a magnetic field induce a current in a coil of wire in that field—a guitar pickup usually features either a single permanent magnet or six magnetic polepieces (one for each string) surrounded by a coil of wire. A popular design invented in the late 1950s is the still widely-used 'humbucking'. or hum-cancelling pickup. This has two wire coils and magnets next to one another, the coils wound in opposite directions and wired in parallel with opposing magnetic poles, so effectively cancelling hum and feedback.

Volume and tone pots

The controls are *potentiometers* ('pots'), variable resistors governing volume and tone settings. The volume pot regulates current output. The tone pot has an added capacitor and regulates the output of high frequencies. When the control knob is turned by the player, the shaft of the pot inside the guitar body pushes a metal protrusion around a circular resistance contact, providing the variable resistance function. The pots are wired to the pickup(s) and the output socket on the face or edge of the guitar body, which feeds the resulting electrical signal out of the guitar to be amplified.

Expansion of tonal control is now supplied on some models by active electronics. A small pre-amplifier is wired into the control circuit and is powered by a battery or by mains. This allows the player to cut or boost tone settings by much wider margins than in normal, 'passive' instruments.

Such sound enhancement, typically providing a cut or boost of ±15 dB at bass or treble frequencies set by the maker, has found the most favour with players of six-string electrics. But a more universal (and practical) advantage of the on-guitar pre-amplifier is its isolation of the instrument from the capacitance (dangerous build-up of current) associated with long cable runs between guitar and amplifier.

Still wider-ranging sounds are demanded by some players, and for this reason some manufacturers are seeking to establish a working link between the electric guitar and the synthesizer. On a synthesizer, a keyboard normally controls voltages to feed

Electronics in Action: Musical Instruments

the oscillators which produce pitch, the filter which provides tonal qualities, and the amplifier which shapes the resulting note(s). The keyboard is thus a control medium, each key providing the relevant voltage.

While the use of the keyboard as a synthesizer controller is assured, the electric guitar can be made to use a synthesizer with control voltages by means of a pitch-to-voltage converter—a device that transfers pitch information into voltage. Problems arise, however, when attempting to follow the subtle inflections and playing characteristics used naturally by guitarists, and to incorporate these nuances into the control signal that feeds the synthesizer. On a keyboard the instrumentalist need only press a key, but the guitarist produces notes and chords by a combination of actions from both left and right hands. Harmonics and overtones from the strings all contribute to the rich sound of the electric guitar.

Synthesizer controller

Early attempts at constructing a guitar controller for the synthesizer side-stepped this problem. For example, the Swedish guitar company Hagstrom, in collaboration with the US electronics firm Ampeg, produced in the late 1970s their Patch 2000 system which completely ignored right-hand picking information and treated the guitarist's fingerboard literally as a keyboard. Each fret became a touch-sensitive switch that provided a control voltage as soon as a string was made to touch it, in much the same way that the keys of a synthesizer keyboard operate.

The Patch system had distinct limitations: the control signal was mono so that only one note could be triggered at a time by the fingerboard; it was not possible to play chords, only single note melody lines; and standard electric guitar sounds were produced by ordinary pickups in the normal fashion, while the synthesizer-driven sound from the fingerboard was added to it only at amplification stage.

Design engineers have now realized that if they are to attract guitarists to synthesizers, the guitar controller must respond to the player's normal range of techniques—both from the left and right hands—on the fingerboard and at the body. Keyboard players, who have been dealing with the intricacies of synthesizers since the late 1960s, are well versed in voltage control; to expect guitarists to play and think like them is unreasonable.

The Roland Corporation in Japan have come up with a partial solution by producing a guitar and control unit combination that will suit the playing styles of guitarists while providing basic synthesizer functions and sounds. Their guitar controllers feature *hexaphonic* six-way pickups in addition to the standard guitar pickups that produce electrical signals amplified in the normal way.

This special pickup collects discrete information from each string and feeds it via a pitch-to-voltage converter to the control

Above Mark Knopfler, lead guitarist of Dire Straits, with his National Steel guitar. *Below* Alembic Flying V bass guitar, and 6/12 string Gibson double-neck bass and guitar.

unit's six voltage-controlled oscillators, one for each string. It is designed to sense the minute variations in pitch information produced by the strings when playing, such as bending a string slightly across the fingerboard or inducing vibrato effects by the regular movement of a finger on a fretted string. Since there is a voltage-controlled oscillator for each string, the guitarist can play chords on the fingerboard and reproduce and treat them through the synthesizer, with various options to tune the pitch of the oscillators against the fundamental pitch of the guitar strings.

The voltage-controlled filter, which produces the tonal quality of the sound, can be controlled on the guitar with knobs for cut-off frequency (the frequency at which attenuation begins) and for *resonance* (the amplitude at the cut-off frequency). Other controls on the guitar include a balance between ordinary guitar sound and synthesizer sound levels, and vibrato on/off touch switches. Vibrato effects are brought about by the low-frequency oscillator which modulates the frequencies of the voltage-controlled oscillators, often at frequencies below normal hearing.

Nevertheless guitar synthesizers still need to be even more responsive to a player's technique, and the future would seem to lie with digital control. Musical instruments using digital control are at present based around a musical keyboard controller, or else an alpha-numeric keyboard with VDU and light-pen for 'drawing' waveforms into a synthesizer. It is probable, however, that digital control will allow greater flexibility when tailoring an instrument's responses to the musician's playing inflections, without being restricted to converting this to some form of control voltage. Synthesizer makers are looking closely at digital techniques for keyboard performance control and touch sensitivity, and will be quick to apply these developments to guitar synthesizers.

Wood substitutes

With one ear on these electronic possibilities, electric guitar makers are also studying the world timber supply. There is no shortage envisaged of the woods most commonly used in guitar manufacture, such as maple (mostly from North America), mahogany (mainly from West Africa and Central and South America), ash (either European or American) and rosewood (from India). Nonetheless, many manufacturers (particularly in the USA) have experimented with new materials as substitutes for wood,

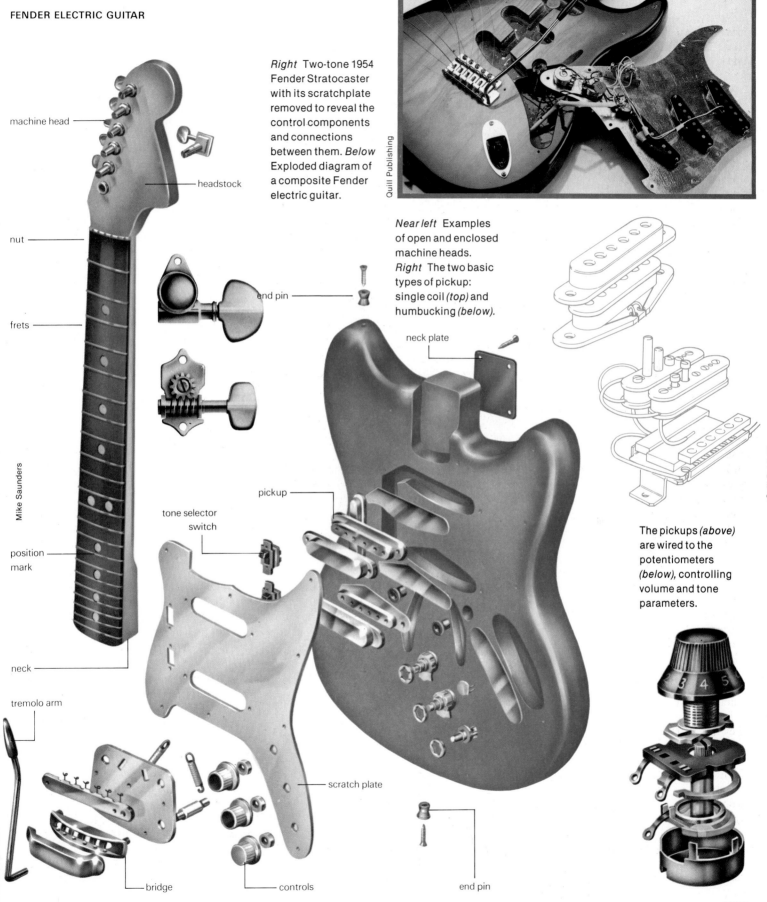

Electronics In Action: Musical Instruments

sometimes gaining cost advantages in the process. Metals and plastics have provided the alternatives, and wide-ranging, often odd-ball, designs have resulted.

Metal has never been entirely suitable in the guitar body or neck—engineers face problems of differing expansion rates in the metallic and wooden sections of the instrument, which in severe cases cause warping and tuning difficulties. Players also complain of the cold feel of metal, particularly when used in the neck of the instrument: on Kramer guitars an aluminium T-section runs the length of the neck, with wood inlays either side in an attempt to provide 'warmth'. The earlier Travis Bean guitars had a solid, machined and rolled billet of Reynolds 6061-T6 aluminium for the neck.

These metallic experiments aimed at improving the sustaining quality of the vibrating strings, based on the theory that a guitar string will vibrate only as long as its mountings will allow. Aluminium has in turn been replaced by brass as the metal most favoured to enhance string vibration and sustain, and it is also widely used in bridges and nuts—the two opposite *stops* for the strings at either end of the electric guitar.

Aluminium has not been completely rejected: in 1981 a very light guitar with a forged one-piece aluminium headstock, neck and body was made by Brunet, and a magnesium neck and cast nickel fingerboard were used by Gauvin, both small makers on the US west coast. For the most part, though, metal in mass-produced electric guitars means steel or brass.

The Kaman Corporation which manufactures aircraft have applied their research into the vibrational and acoustical physics of helicopter rotor blades to the acoustic guitar. Ovation, part of the Kaman Corporation, produced as a result what is perhaps the most revolutionary use of plastics for guitar bodies, and the only notable recent development of any kind in the making of acoustic guitars. Ovation's famous guitar has an unconventional round-backed body made from a glassfibre substance, patented as 'Lyrachord', which produces a high level of crisp sound projection. The more recent introduction of a bridge-mounted transducer, connected to pre-amplifier offering volume control via a potentiometer, brought the electric acoustic guitar into being.

Ovation also make more conventional solid-bodied electric guitars, and in this area

ROLAND G303 GUITAR

Considerable research has been devoted to producing a suitable synthesizer for guitarists. Some require a custom-made guitar, others use a special pickup that can be attached to an existing instrument. The Roland GR300 synthesizer combined with the Roland G303 polyphonic guitar *(right)* is very popular.

Left Andy Summers of Police using a G303 polyphonic guitar. Its complementary synthesizer unit GR300, in front of him, includes foot controls although frequently used controls are located on the guitar itself. *Right* The body construction of the Gibson Sonex guitar is a sandwich of Resonwood surrounding an inner core that is made of maple wood.

they have again exploited new materials and new combinations of materials. They firmly maintain that the sound quality of the electric guitar is largely determined by the stiffness and overall rigidity of its construction, especially the relationship between the nut, bridge and frets, and their UK II guitar incorporates these qualities.

The best known of the older American guitar manufacturers is probably the long-established Gibson Company, but tradition has not deterred it from successful involvement with plastic bodies. Their new Sonex guitar has a body consisting of a central core of conventional maple sandwiched between outer layers of a glassfibre-like wood-resin material. The company have announced that variations on this basic combination, known as Resonwood, will soon be marketed.

While the Gibson instruments have been designed to benefit from the lower costs of

THE OVATION ACOUSTIC GUITAR

Exponents of the Ovation guitar include John Denver and John Williams.

Ovation electric acoustic guitars are based on two new design concepts: the bowl-like back *(below)* and a bridge-mounted pickup system.

Below Mick Jagger with his Gibson SG guitar. *Bottom* A corner of the Mississippi-based Peavey plant, where modern technology has been adapted to the manufacture of guitars.

new materials just as much as from their superior sound, other US companies are investigating the properties of materials applicable to expensive top-line instruments.

Such a company is Alembic, whose latest guitar has a neck made of carbon fibre processed under high pressures and temperatures. Although much lighter in weight when compared to normal neck woods, the carbon fibre provides a greater degree of stiffness and also better dimensional stability—no expansion or contraction occurs with changes in temperature or humidity. Alembic additionally claim that the resonant frequency of carbon fibre is higher than any of the fundamental frequencies produced from the fingerboard, which means that there is no interaction with the instrument's sonic response and accuracy.

Peavey, another US manufacturer, developed a triple-density plastic called Sustantite for guitar use. This had a central core, an inner layer of 'blown' plastic for shaping and an outer plastic skin over the top. It has been abandoned because, incredibly, the cost of producing body blanks from oil-based plastics has now caught up with using maple as a raw material. The same company's development of a plastic neck was found to be less successful, due to the great stresses incumbent on the neck.

Automated production

Nevertheless modern technology still plays a part in the Peavey factory, purpose-built in 1973. Computers are used to control various functions, such as the numerically controlled West German neck-cutter developed from a furniture leg-turning machine because of its versatility in cutting asymmetrical wooden shapes. Buffing the necks is performed at great speed by a robot and, elsewhere in the factory, bodies are cut, routed and shaped on numerically controlled devices.

All these techniques are not revolutionary in general industry, but until recently they have been rare in guitar manufacture. Japan, a country well-known for its use of computers and robots, is an important and successful producer of electric guitars. At the Yamaha factory, for example, a totally computer-controlled digital router is used for carving and shaping guitar bodies, simultaneously routing pickup and control cavities. In fact throughout the industry, whether in Japan or the USA, manufacturers are using automated techniques for routine work in guitar making, while retaining craftsmen and musicians for the many essential tasks in the design, production and marketing of their products.

Medical Science: Neurology

The big headache

Migraine remains one of the great mysteries of medical science, despite the fact that its complex symptoms were identified as early as 400 BC. Finance for research has always been much more limited for migraine than for other similarly widespread and debilitating diseases, perhaps because it is not fatal. But even so, migraine currently affects around one in ten people worldwide, and it costs Britain alone £20 million a year in lost working days.

Many factors contribute to the difficulty of preventing or curing migraine, not the least being that most people suffer at home and that a wide variety of manifestations and *triggers* (things which bring on the complaint) have been described by *migraineurs*, the migraine sufferers. Some confusion surrounds what a migraine actually is, partly because of the variable forms of a true migraine and partly because some people mistakenly think of it as any severe headache. In fact there are three basic types: classical, common and complicated migraine.

The latter is rare and can include a temporary paralysis of the limbs and loss of external occular movement.

Hallucinations

Classical migraine, on the other hand, is characterized by transient visual disturbances —medically termed *scintillating scotoma*. Scotoma are the longer lasting and more elaborate hallucinations within the visual field suffered as the attack progresses, while scintillating scotoma is the characteristic flickering of luminous migraine spectra. The period in which the hallucinations take place is known as the aura, a term long used for the hallucinations that occur at the beginning of an epileptic fit.

Some migraineurs experience what can only be described as a change in perspective at the start of an attack, followed by an aura in which distance and space appear to be distorted. Those able to identify this can take medication immediately and sometimes reduce the severity of the migraine or banish it altogether. But once the attack becomes more advanced, gastro-intestinal activity decreases significantly and the effectiveness of medication is reduced.

Classical migraine can also include abdominal disturbances which cause *anorexia*

This woman exhibits three of the most common symptoms of a migraine attack: one-sided headache, nausea and visual disturbances known as scintillating scotoma. Exhaustion— in her case from shopping—often triggers migraine attacks in habitual sufferers.

(loss of appetite), nausea and, in a severe attack, vomiting. *Photophobia* (dislike of bright lights) is common, as is difficulty in focusing. Some migraineurs find that bright lights—especially sunlight flashing through trees or a flickering television set—precipitate an attack. Sensory or motor *prodromes* (symptoms signifying attack) include numbness and clumsiness.

Common migraine, as distinct from other types of headache, does not have easily identifiable prodromes, but it is associated with any two of the following factors: nausea, *unilateral* (one-sided) pain, or a family history of similar headache. Unilateral headache is a feature of all migraine and it was this that led Galen to name it *hemicrania*—half-skull—in the second century AD. This led in turn to emigranea, migranea and migraine.

Variables of migraine include migrainous neuralgia, where attacks come in 'clusters' in the early morning with pain invariably behind the same eye, and tension headaches. The latter arise from excessive contraction of head muscles that include those used for chewing and biting: as these muscles lie on the surface of the skull, there may be local tenderness. The pain is usually felt as a pressure sensation on top of the head or in the forehead and it is usually continuous, lasting all day. Tension headache differs from migraine in that there is only headache and

Above Migraineurs are often sensitive to flickering lights, movement and noise, so a disco strobe show is an obvious trigger for attacks. *Right* Migraineurs have experienced or exhibited all these symptoms in varying combinations during attacks. *Left* One recent discovery in migraine research: migraineurs' blood reacts abnormally to an increase in amine levels caused either by stress or eating certain foods.

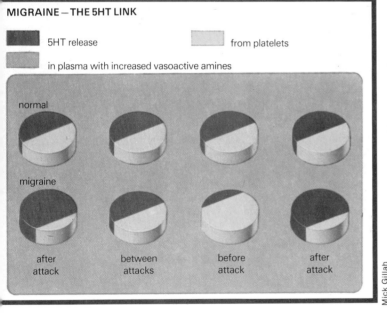

MIGRAINE—THE 5HT LINK

- 5HT release
- from platelets
- in plasma with increased vasoactive amines

normal / migraine — after attack, between attacks, before attack, after attack

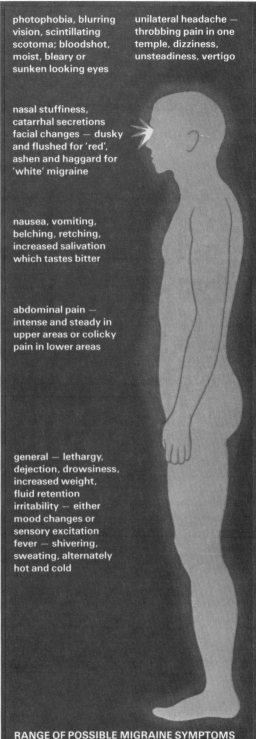

- photophobia, blurring vision, scintillating scotoma; bloodshot, moist, bleary or sunken looking eyes
- unilateral headache — throbbing pain in one temple, dizziness, unsteadiness, vertigo
- nasal stuffiness, catarrhal secretions facial changes — dusky and flushed for 'red', ashen and haggard for 'white' migraine
- nausea, vomiting, belching, retching, increased salivation which tastes bitter
- abdominal pain — intense and steady in upper areas or colicky pain in lower areas
- general — lethargy, dejection, drowsiness, increased weight, fluid retention irritability — either mood changes or sensory excitation fever — shivering, sweating, alternately hot and cold

RANGE OF POSSIBLE MIGRAINE SYMPTOMS

muscle tension, and not scotoma or nausea.

Migraine has for many years been believed to originate with an element or elements of the diet, hormonal fluctuations experienced by menstruating women, or by stress, either physical or mental. These are generally accepted as resulting in constriction of the arteries near the scalp and inside the head, producing migraine. However, it has been demonstrated recently in Copenhagen, Denmark, that this constriction—traditionally thought to be the cause of migraine rather than an accompaniment—does not take place in common migraine.

The greatest problems with migraine research lie in the unavailability of the patient during an attack and the great variety of triggers. The first problem has to some small extent been alleviated by the establishment in recent years of migraine clinics for research and on-the-spot treatment, monitoring and analysis. Clinics have been set up in Italy, Belgium, the Netherlands, Norway and Britain, of which the Princess Margaret Clinic in London is notable.

Of the food triggers, cheese, chocolate and citrus fruits are well known, but the list is much longer: alcohol, particularly red wine; fried fatty foods; vegetables; tea; coffee; meat, particularly pork; seafood; wheat; foods containing monosodium glutamate —which has resulted in the term 'the Chinese restaurant syndrome'. All these foods have one thing in common: they contain either amines or organic chemical compounds that contain the amine group in their structure. It is these amines that have come to be associated with migraine attack.

The chief vasoactive amines implicated in migraine attack are adrenaline, dopamine, noradrenaline, octopamine, histamine, beta-phenylethylamine, tyramine, and 5-hydroxtryptamine (5HT or serotonin). Citrus fruits

Medical Science: Neurology

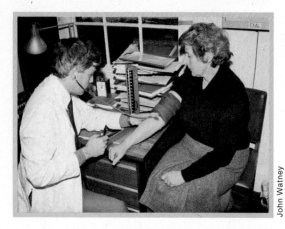

Treatment at the City of London Migraine Clinic starts with a complete medical examination—including a blood pressure check—and recording the medical history.

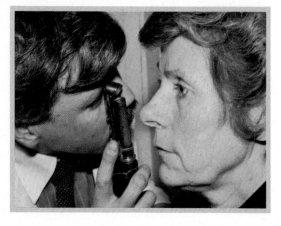

The backs of the eyes are examined to assess the state of the arteries and inspect the nerve to the brain, as well as to check for any signs of raised intercranial pressure.

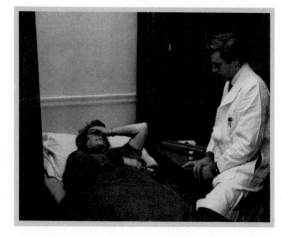

If patients have attacks while in the clinic, they rest in a darkened, quiet room—migraineurs are often sensitive to bright light and noise during one of their attacks.

contain octopamine and these have been implicated in migraine attacks for years, particularly when taken on an empty stomach.

Dr Edda Hanington working with the Haemotology Department of St Bartholomew's Hospital, London, observed that the tyramine found in cheese affects the blood vessels in the head, giving rise to headache, certainly in classical migraine. Chocolate contains another amine, betaphenylethylamine, in large quantities. Wines contain histamine, another powerful blood vessel dilator (as can be seen in an inebriate's flushed face).

Dr Hanington has suggested that migraineurs lack the enzymes which break down these amines. Increased permeability of the gut would mean that the amines could circulate the body freely, and ultimately affect the blood vessels, provoking an attack. This hypothesis is currently under investigation at the Princess Margaret Clinic.

Adrenaline and noradrenaline, the catecholamines, are the chief substances that are released in the body during periods of stress. When large quantities are released, a migraine attack can result. Likewise, histamine is produced in increased amounts in the body in response to shock and allergy.

Suggestible complaint

There are many problems surrounding the isolation of which particular food a migraineur cannot tolerate, not the least of which is that migraine appears to be a suggestible complaint. Dr Hanington conducted a survey of 500 migraineurs and found that 74 per cent considered chocolate precipitated headache. However, when she gave chocolate to one group of volunteers and an identical-looking placebo to another, about 30 per cent of the second group reacted with headaches.

One of the methods under current investigation, the radioallergosorbent test (RAST), avoids this problem. The test relies on an antibody-antigen reaction: a sample of the migraineur's blood is mixed with the cultured antibody directed against an antibody to a particular food; if an antigen reaction takes place, there would be an association between this sensitivity and a migraine attack.

It is now thought that 'dietary' migraine may be the problem for about 70 per cent of patients, but that the problem is either one of food allergy or the absence of the necessary enzymes in the intestine to cope with the ingestion of a particular food.

Other sorts of allergy in migraine current-

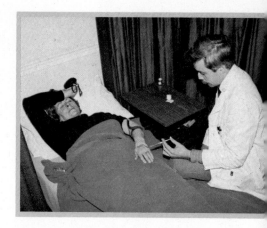

Oral medication can be taken if the patient is nauseous, but if vomiting has started an injection of the anti-nausea drug Metaclopramide is given.

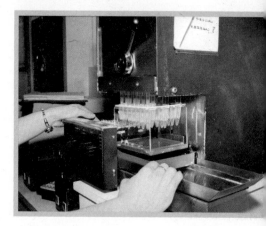

After taking the medication—Ergotamine—blood is diluted in the 24 tubes of the Kone Analyzer and drug levels assessed by a delicate radio-isotope method.

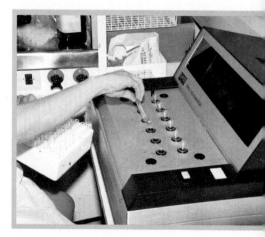

The Multi-Detector Gamma Counter is also used to check drug levels. The highest levels of Ergotamine were found in cases where the drug was taken as a suppository.

ly under investigation include things which trigger hay fever and asthma, such as dust. An efficient means of skin testing is being researched, as is *immunotherapy* (a programme of desensitization) both for food and for other allergens.

One of the most significant recent discoveries in migraine research has been the observation that migraineurs show an increased ability in blood *clotting*—the aggregation of blood platelets. And, furthermore, the migraineur's blood contains a releasing factor during an attack which releases 5HT from platelets obtained during a headache-free period, causing blood vessels in the area of their release to constrict.

The fact that blood vessels constrict during a migraine attack has been known for many years, but the exact role of 5HT in migraine is still not fully understood. However, while 5HT levels in normal and migrainous subjects are the same, these levels sometimes rise three times above the normal just before a headache starts.

It is also interesting that migraineurs re-

Above Sunlight flashing through trees can precipitate migraine. *Top right* The normal pattern of platelets in blood.
Below right Platelets aggregating during a migraine attack. This may be the root cause of strokes—to which migraineurs are more prone than non-sufferers.

quire only very low dosages of lysergic acid (LSD)—an ergot compound—to achieve a 'trip'. This seems to indicate that the sensitivity of their brainstems to 5HT is higher. And, interestingly, ergot compounds have a tradition of use in migraine treatment.

Although one in ten people suffer migraine attacks, the incidence is only one per cent or less in five-year-olds, rising to 13 per cent of boys and 14 per cent of girls in early adolescence, rising again to 19 per cent of adult males and nearly 26 per cent of adult females. Of all women, about 40 per cent experience one attack in their lifetime. The differential between female and male migraineurs continues through the menstruating years and then gradually decreases.

Because of this difference, medical research has investigated the association between migraine and hormonal fluctuations. It has been shown that about 20 to 25 per cent of female migraineurs regularly experience an attack premenstrually, or occasionally at mid-cycle. The particular hormones are oestrogen and progesterone, which are at their lowest levels at the onset of menstruation. It has already been found that the oestrogen level in migraineurs tends to remain somewhat higher than average, that migraine often clears up during pregnancy when substantial changes in the balance of hormones take place, and that the contraceptive pill—which contains oestrogen—can increase the incidence and severity of attacks.

Migraine in children is comparatively rare —and even then not always diagnosed. Many child migraineurs complain of feeling

Medical Science: Neurology

sick and giddy but rarely experience the aura of classical migraine. They do, however, frequently feel travel-sick—especially if over-excited or on a long journey without stops for proper meals—and people who only begin to suffer as adults often recall that they were travel-sick as children.

Headache is not normally associated with the teeth, but experimental research now in hand may show that there is a link between *bruxus* (night-time tooth grinding), *bruxomania* (day-time grinding), clenching of the jaw and migraine. Grinding and clenching are well known as automatic reactions to stress but the British Dental Migraine Study Group believes they may also be connected with a faulty *bite*—the position of jaw and teeth when at rest. Bruxus may be the reason why many migraineurs suffer on waking and may be caused by a faulty bite resulting from the removal of some back or eye teeth. The end result of a faulty bite can be sore jaw joints and *masseter muscle* (at the side of the face) which in turn give rise to the one-sided headache characteristic of migraine.

Acupuncture has often been dismissed as a psychological cure—and it is unfortunately true that one in four migraineurs initially benefit from almost any 'new' therapy—but acupuncture has proved successful in some cases. So has biofeedback, in which the patient controls his temperature.

One of the most important discoveries so far—made by means of electrical testing through the retina to the cerebral cortex—is that classical migraineurs have a delay in the impulses travelling from the front to the back of the brain. It is thought that there may be an abnormality of the brainstem in migraineurs and that this may be associated with 5HT neurones.

Smoking and migraines

Also of interest are the discoveries that 80 per cent of migraineurs are smokers, while for the population as a whole the figure is closer to 45 per cent; that epilepsy is six times more frequent in migraineurs; that there is a disproportionate number of migraineurs among hangover sufferers; and that migraineurs generally have higher sensitivity to pain—in tests involving tolerance to *white noise* (jumbled noise like the 'whoosh' emitted by television sets when programmes are not being transmitted), migraineurs found about 90 decibels tolerable, whereas the population as a whole could take another 20 decibels on average.

Many migraineurs hold their heads to the side or forwards—relaxation exercises can help by easing pressure on muscles. The Alexander Technique *(below)* changes posture and movement habits, using a complicated learning process. It distinguishes 'bad use' of the body —(2), (3) and (5)—from 'good use'—(1) and (4).

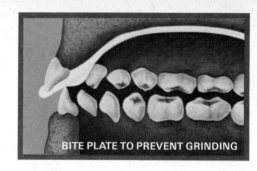

BITE PLATE TO PREVENT GRINDING

One school of thought on migraine causes blames a faulty bite for tooth grinding, leading to headaches. Bite plates such as the modified Hawley *(above)* have helped some cases. *Below* Scintillating scotoma from inside—a migraineur's painting.

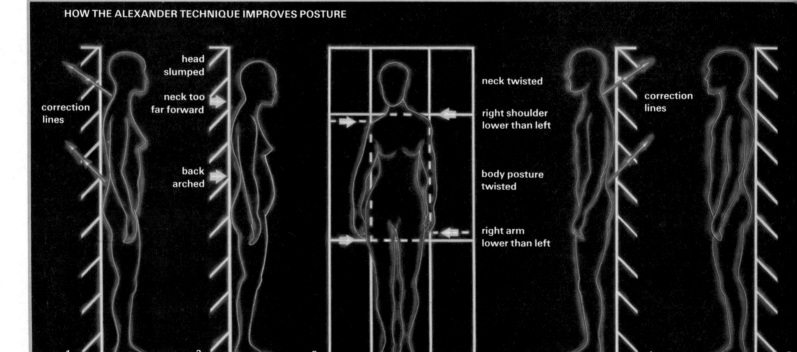

HOW THE ALEXANDER TECHNIQUE IMPROVES POSTURE

Frontiers: Space

Venus: inferno in disguise

The planet Venus—15 times more brilliant than the brightest star—can shine brighter in Earth's sky than any object, except the Sun and Moon. At its most brilliant, Venus appears to hang like a lamp in the morning or evening twilight—hence 'Morning star' or 'Evening star'—but until 20 years ago, astronomers knew nothing about its cloud-shrouded surface. Now, a combination of powerful radar instruments on Earth and a long series of space probes to Venus have unveiled the mystery planet.

Named after the Roman goddess of love, Venus is the Earth's nearest neighbour. Although roughly the same size as Earth, it is generally much smoother despite being broken in one place by a mountain peak higher than Everest. And it is blanketed by an atmosphere of carbon dioxide, 90 times denser than the Earth's, which raises its surface heat above the melting point of lead.

The planet's brilliance is due to its size —almost the Earth's twin, with a diameter of 12,100 km (7,500 miles)—its close approach to Earth and its enveloping, brilliant, yellow-white clouds. These clouds have prevented even the most powerful Earth-based telescopes from seeing Venus's surface, so the nearest planet remained a mystery up until the early 1960s.

Although clouds can effectively block off light, they have no effect on radio waves. So a radio signal from Earth to Venus will penetrate the clouds and be reflected back by the planet's solid surface. Analysis of this radio 'echo' (called *radar astronomy*) can then reveal much about the world beneath the clouds. In 1961, radar astronomers first developed the powerful transmitters and sensitive receivers necessary for producing detectable echoes over the 41,400,000 km (25,709,000 miles) between Earth and Venus.

The following year, these same astronomers achieved sufficient sensitivity to find the rotation rate of Venus. The results showed that Venus rotates much more slowly than the Earth, turning once in 253 days. Even more surprising, Venus does not rotate from west to east, like the Earth, but from

Left In this rare view the disc above Earth is the Moon, above this is Venus, and above this is the comet Kahoutek. **Above** Face of Venus.

east to west. Besides the tiny, highly tilted Pluto, and Uranus—which spins virtually on its side—Venus is the only planet to spin in this *retrograde* (or backwards) sense, opposite in direction to its orbit about the Sun.

Venus's peculiar rotation has an odd—and inconvenient—effect. When Earth and Venus are at their closest, and best placed for radar observations, Venus has completed exactly four rotations since its previous closest approach—so the same part of the planet is facing Earth. This limits radar astronomers to studying only one face of the planet.

The early radar echoes showed two highly reflecting areas on Venus, named Alpha and Beta. Later radar results have found that Alpha consists of parallel mountain ranges, with a ring-shaped feature—named Eve—to the south. Eve is probably a large crater, and its central peak has been chosen to mark the zero of longitude—'Greenwich' of Venus.

Beta consists of two conical peaks close together. One is 5,000 m (16,400 ft) high, and is topped by a shallow dip—probably a huge volcano. Earth-based radar also showed a large peak near Venus's North Pole, named Maxwell after the Scottish physicist James Clerk Maxwell whose theory of electromagnetism first predicted radio waves.

Probing Venus

Despite great technological efforts, even the radar results have had limited use. They cover only part of the planet, and despite their ability to resolve small details, they cannot reveal heights accurately. Also, radar gives away nothing about Venus's atmosphere, its winds and clouds; most of what is known in these areas has come from US and Soviet space probes to the planet.

The US probe Mariner 2 was the first to reach another planet. It flew past Venus at a distance of 35,000 km (22,000 miles) on 14 December 1962. Its instruments measured the planet's temperature beneath the clouds and showed, to the surprise of most astronomers, that Venus is extremely hot: a surface temperature of 470°C makes the planet even hotter than the most scorched parts of Mercury. A later US probe, Mariner 10, took the first close-up photographs of Venus in February 1974, as it flew by on its way to Mercury. Its pictures revealed the surprising fact that Venus's clouds travel around the planet 60 times in the period that the planet itself takes to turn once.

Meanwhile, the USSR had started launching a series of Venus probes to penetrate the atmosphere. The first three successful probes—Venus 4, 5, and 6—transmitted data as they descended on parachutes, but they were crushed by intense atmospheric pressure before they reached the surface. The highly reinforced probes, Venus 7 and 8, landed on Venus in 1970 and 1972 and relayed data showing that the atmospheric pressure is about 90 times Earth's.

In 1975, the USSR pulled off an impressive technological feat by landing two craft with cameras. These returned the first (and so far the only) photographs of the surface of Venus. Venus 9 recorded a landscape consisting of angular, jagged rocks but Venus 10 landed in a region of flatter, 'weathered' rock outcrops; surprisingly, the surface was bright under the cloud pall. However, the real breakthrough in knowledge about Venus has come from the

COMPARATIVE SIZE AND ROTATION OF EARTH AND VENUS

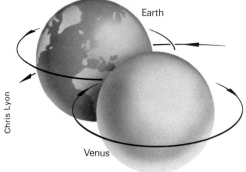

THE ATMOSPHERE OF VENUS

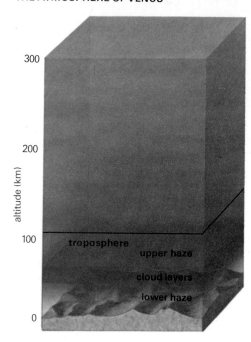

Above The atmosphere of Venus is divided into cloud layers of varying density.
Left Earth and Venus are comparable in size, but they rotate in opposite directions.

THE SURFACE OF VENUS

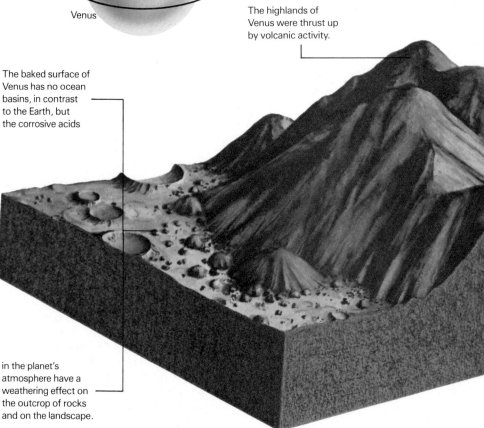

The highlands of Venus were thrust up by volcanic activity.

The baked surface of Venus has no ocean basins, in contrast to the Earth, but the corrosive acids in the planet's atmosphere have a weathering effect on the outcrop of rocks and on the landscape.

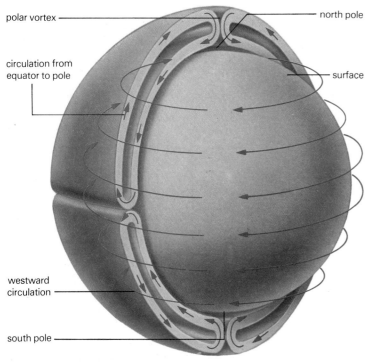

WINDS IN VENUS'S LOWER ATMOSPHERE

- polar vortex
- circulation from equator to pole
- westward circulation
- south pole
- north pole
- surface

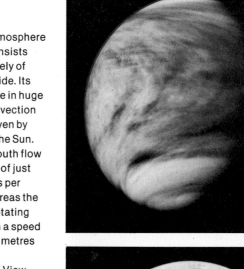

Left The atmosphere of Venus consists almost entirely of carbon dioxide. Its clouds rotate in huge swirling convection currents driven by the heat of the Sun. The north-south flow has a speed of just a few metres per second whereas the westward rotating winds attain a speed of some 100 metres per second.

Above right View of Venus taken with Orbiter's ultraviolet instrumentation during its orbit in January 1979.

Right Venus's 'continents' in an artist's impression based on radar data from the Pioneer Orbiter. Three continent-sized high land masses can be distinguished.

Slow winds in the lower atmosphere disturb the surface only lightly.

The lowlands consist mainly of rolling planes covering about 60% of the surface.

Despite huge rift valleys on Venus, the planet has no history of drifting land masses that caused similar features on Earth.

Left Aphrodite Terra is one of the Venusian highland areas, raised several kilometres above the average level. This mountainous region stretches along the equator for almost half the planet's circumference. Aphrodite Terra's central area is split apart by huge rift valleys, some 3,000 m (10,000 ft) deep and hundreds of kilometres wide.

US Pioneer expedition to Venus in 1978.

One of the two spacecraft comprising the expedition consisted of a large 'bus' carrying four smaller probes. These separated on arrival at Venus and descended at different locations to study the atmosphere at various latitudes and on the 'day' and 'night' sides of the planet. The second spacecraft was an orbiter. Since 4 December 1978, it has been travelling around Venus in an elongated, tilted orbit which takes it from only 150 km (90 miles) above the surface to a far point of 66,000 km (41,000 miles)—5.5 Venus diameters. The Orbiter has taken photographs of the clouds to study their motions, and investigated their infra-red emission. It also carries a small radar set, which cannot 'see' details finer than 100 km across (ten times less detail than the best Earth-based radar) but which is so close to Venus that it can measure heights with great accuracy.

The Pioneer Orbiter's radar measurements have shown the shape of Venus's land masses below the clouds. In general, slow-rotating Venus is an almost perfect sphere—unlike the Earth, which has a distinct bulge around the equator because of its rapid spin. The

Frontiers: Space

most obvious difference between maps of Venus and the Earth is that the baked surface of Venus has no oceans.

Venus, on the other hand, consists mainly of rolling plains at an intermediate level. These plains cover about 60 per cent of the planet's surface, but only 16 per cent of Venus is in the form of lowland basins. The highlands of Venus lie within three distinct regions, each much smaller in area than the main continents of the Earth.

Geologists had expected that Venus, being the Earth's twin in size, should have a similar amount of interior heat and so bear a similar pattern of *plate tectonics*—the recent formulation of the old idea of continental drift. But Venus is different. Indeed, the lack of continental drift on Earth's twin planet is one of Venus's greatest mysteries.

The thick atmosphere of Venus differs greatly from the familiar air of planet Earth, with its 21 per cent oxygen, 78 per cent nitrogen and smaller amounts of argon, water vapour and carbon dioxide. Venus's atmosphere comprises 96 per cent carbon dioxide, and most of the remaining 4 per cent is nitrogen. There is virtually no free oxygen, and only 0.01 per cent water vapour, with a roughly equal amount of sulphur dioxide and argon.

Higher up in the atmosphere, water vapour and sulphur oxides combine to form great banks of sulphuric acid droplets. These appear as the beautifully shining cloud layers of Venus but are unlike the soft, water-bearing clouds of Earth. The densest clouds lie at a height of 50 km—five times higher than the highest clouds on Earth. This layer is about 3 km thick, and above and below are more extensive layers of acid mist. From these clouds, drops of sulphuric acid probably fall to corrode the planet's surface.

Puzzling heat

Although Venus is closer than Earth to the Sun its bright cloud and haze layers reflect 80 per cent of sunlight back into space, so the amount of solar energy actually reaching the surface is only about two-thirds that reaching the surface of the Earth. At first sight this seems paradoxical because Venus's surface is at such a high temperature. The major difference between the two planets, however, is not in the energy that they receive from the Sun but in the amount they radiate back to space. The Earth's atmosphere lets infra-red radiation escape with almost the same ease as it lets the light in, so Earth has a temperature only slightly above the value expected for a planet that is situated its distance from the Sun.

The atmosphere of Venus, however, has a composition which makes it much more difficult for infra-red radiation to escape. This ability to trap solar energy is the *greenhouse effect*. Venus's greenhouse effect is due mainly to the infra-red absorption by carbon dioxide in its atmosphere, but the small amounts of water vapour contribute about one-quarter of the effect, and the clouds and haze about one-sixth. Together they raise the planet's temperature more than 400°C above that expected for a world at Venus's distance away from the Sun.

The Pioneer results have also helped to explain the circulation of Venus's atmosphere and clouds. The main problem has been to explain the rapid rotation period of the cloud layers. It means that the clouds are speeding around Venus at 362 km/h (225 mph) relative to the planet's surface.

Less than a quarter of Venus's atmospheric gases lie at this height, where the high winds

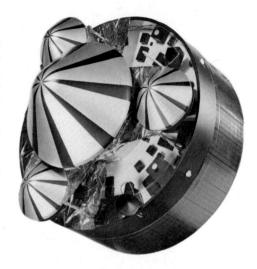

The USSR landed a craft *(right)* on Venus in 1978 and took the first (and so far the only) photographs of the planet's surface *(below)*. But the real breakthrough in knowledge came in 1978 when the US Pioneer space probe *(above)* entered the Venusian atmosphere.

occur. The bulk of the atmosphere (lying at lower altitudes) has low wind speeds, less than 20 km/h relative to Venus's surface. This massive lower atmosphere is rotating with the planet. But as gases rise into the upper atmosphere, they collide at increasing heights with less massive 'parcels' of successively more rarified gases, which are propelled faster and faster. Eventually, the rising gases can generate the rapidly swirling winds in and above the cloud layers.

As well as this fast motion around the planet, Venus's atmosphere has a slower circulation from equator to pole. Any individual 'parcel' of gas rises at the equator, due to the Sun's heating effect, and as it whirls rapidly around Venus it also spirals gradually towards one of the poles. Eventually, it travels down again, through one of a pair of vortices at each pole. These vortices show up clearly in infra-red photographs from the Pioneer Orbiter as regions that are hotter than the rest of the clouds. Through the vortex centres, lower layers of the atmosphere are visible, and these are warmer than the general cloud tops.

For twenty years, Venus has produced surprise after surprise to show it is Earth's twin planet only in size. Some of these surprises are still inexplicable, such as the slow retrograde rotation, but astronomers are beginning to explain other differences between Venus and Earth as an effect of their slightly different distances from the Sun dur-

Though Pioneer's probe is unable to see detail under 100 km across—worse than the largest Earth-based radar *(below)*—it measures height with more accuracy.

ing the crucial early phases of their lives.

The Pioneer probes have provided some important clues for astronomers studying the details of planet formation—finding, for example, 75 times as much original argon on Venus as on Earth. But the composition of the two worlds must have been similar in common elements such as carbon, oxygen, silicon and iron. As the two rocky planets condensed into globes, their volcanoes should have exhaled roughly similar amounts of gases, to cloak them in chemically almost-identical atmospheres.

To judge by terrestrial volcanoes today, the bulk of these gases would have been water vapour and carbon dioxide, with smaller amounts of nitrogen and sulphur-oxide gases. At the Earth's distance from the Sun, the planet's temperature was sufficiently low for water vapour to condense and rain down into the great oceans. The water dissolved some of the carbon dioxide and, once in solution, this gas reacted with existing rocks to produce carbonate rocks. Most of the other gases in the atmosphere were washed out too, leaving mainly the almost-insoluble nitrogen. Eventually, the remaining carbon dioxide was converted to oxygen by photosynthesis in plants.

Venus, however, was too close to the Sun for its temperature to drop below the boiling point of water. The carbon dioxide stayed in the atmosphere, blanketing the Sun's heat further, so that Venus's temperature soared in an irreversible, runaway greenhouse effect. Sulphur-oxide gases rose upwards and condensed into the great acid clouds. Nitrogen remained, but only as a minor proportion of the carbon dioxide.

This simple, elegant theory leaves one major problem. Originally, Venus must have been shrouded in as much water vapour as the Earth has liquid water in its oceans. Yet today, water is only a minor part of Venus's atmosphere, its molecules outnumbered ten thousand to one by those of carbon dioxide. So where has Venus's water gone?

Most likely, water vapour in the upper atmosphere has been split up by the Sun's ultraviolet radiation, and the hydrogen atoms could then have reacted with Venus's surface, and become locked up in the rocks.

Hellish heat

Although Venus is so inhospitable today, it might in fact be improving from an even worse past. The original atmosphere, rich in water vapour as well as carbon dioxide, should have trapped even more sunlight and heat than today. Some calculations suggest that the surface temperature would then have been 1,500°C—white-hot incandescence, hotter than a blast furnace. While water vapour was condensing on Earth to form the great water oceans, Venus might have been enveloped in molten rock.

Further unmanned probes are planned to investigate the intriguing planet Venus further. These include a joint Soviet-French mission for the mid-1980s, and an ambitious US probe with an extremely detailed radar mapper for about 1990. But there can be no manned expeditions to the planet—despite its alluring name. An astronaut stepping out unprotected on the planet's surface would survive for no more than a fraction of a second as he or she were simultaneously roasted, suffocated, crushed and corroded.

Medical Science: Virology

Cold cures: not to be sneezed at

While devastating viral diseases such as smallpox and polio have been virtually eradicated, the common cold—the most widespread and trivial viral infection—still eludes prevention and cure. In the UK, each person can expect to get between two and five colds a year, and in the USA, two hundred million days of work and school are lost through colds each year. However, recent advances in our understanding of viruses and the body's immune system offer promising prospects for the future.

Traditionally, the common cold has been defined in clinical terms and distinguished from influenza or allergy, for example, by what is known as a *syndrome*—the collection of apparently related subjective symptoms reported by the patient and objective signs detected by the investigator. Among the early subjective symptoms are feeling 'off colour' and stuffy, headaches and muscle aches. Later, the sufferer may experience fever, chills, sore throat and sore eyes.

Objective signs of a cold include excessive nasal discharge, sneezing, coughing, red and swollen nasal lining, raised temperature, red throat and red and watery eyes. These signs and symptoms are not precise phenomena, and all need not be experienced before diagnosis can be made, but each represents part of a continuous spectrum.

The course of the common cold typically extends over one week, although this is subject to great variation. Similarly, the severity of the symptoms ranges from a mild snuffle lasting one or two days to a profuse nasal discharge associated with sore throat, cough, headache and fever, necessitating several days' bed rest and lasting two to three weeks overall.

Occasionally, the damage done by the initial infection allows a second—often bacterial—infection to get a foothold, resulting in one of the so-called complications of the common cold. These include sinusitus and ear and chest infections. More importantly, a cold may worsen an existing

COMMON COLD VIRUSES IN ORDER OF PREVALENCE

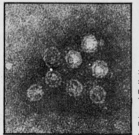
rhinovirus

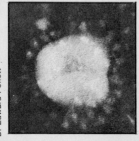
coronavirus

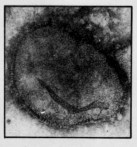
respiratory virus

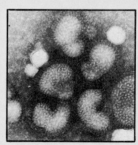
paramyxovirus

adenovirus

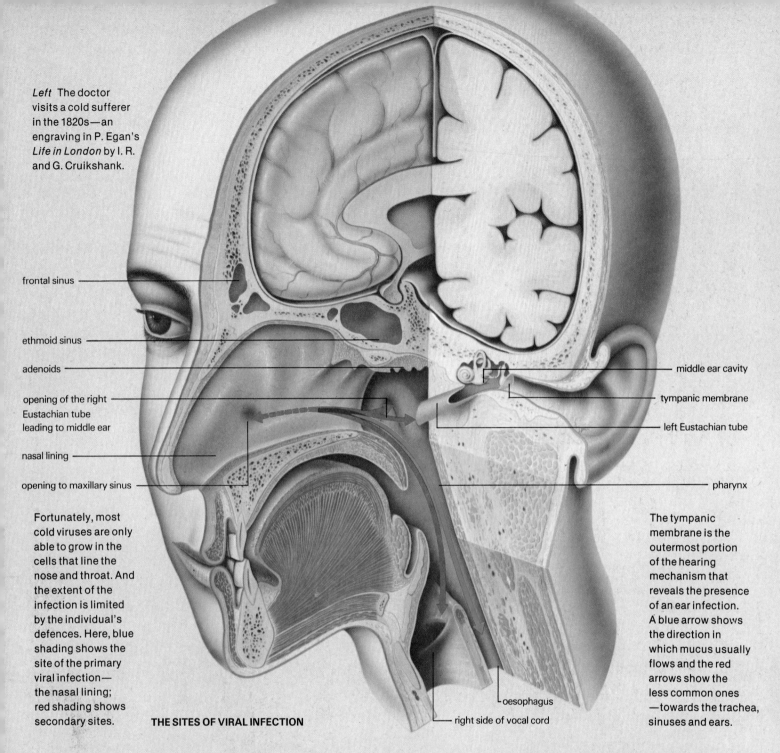

Left The doctor visits a cold sufferer in the 1820s—an engraving in P. Egan's *Life in London* by I. R. and G. Cruikshank.

Fortunately, most cold viruses are only able to grow in the cells that line the nose and throat. And the extent of the infection is limited by the individual's defences. Here, blue shading shows the site of the primary viral infection— the nasal lining; red shading shows secondary sites.

THE SITES OF VIRAL INFECTION

The tympanic membrane is the outermost portion of the hearing mechanism that reveals the presence of an ear infection. A blue arrow shows the direction in which mucus usually flows and the red arrows show the less common ones —towards the trachea, sinuses and ears.

illness such as asthma or chronic bronchitis.

There are two very important reasons why the features of the common cold vary so greatly from person to person. First, the clinical manifestations of any infection are the product of the properties of the infecting organism and the response of the patient and his or her defence system to these properties. Hence, an individual with a relatively poor defence system may suffer worse symptoms than someone with a good defence system.

Secondly, although the common cold is treated as a single entity, in reality infections with at least five different groups of viruses comprised of about 150 different virus types can produce the clinical syndrome of the common cold. Furthermore, it is often not the only disease caused by these individual virus types. In fact, it can be rare.

Viruses had been suspected as the cause of the common cold ever since 1914 when W. Kruse, a German doctor, filtered bacteria from the nasal secretions of his cold-suffering assistant and showed that the filtrate could cause colds if instilled into volunteers' noses.

However, it was not until the early 1960s that any particular virus could be definitely said to cause the common cold. Much of the early pioneering work was performed by Sir Christopher Andrewes and his colleagues at the Common Cold Research Unit near Salisbury, UK, where it was possible to study the transmission of colds amongst volunteers in isolation and isolate and characterize many cold viruses.

Research was complicated by the fact that viruses are mere packages of information —needing the biochemical processes of cells in order to reproduce themselves—so they must be grown within animal cells. This can be done in living animals, but in order to study the chemistry and structure of viruses

Medical Science: Virology

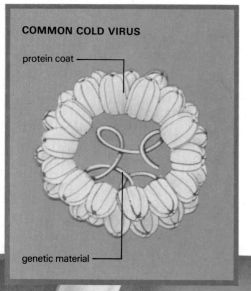

it is almost essential to grow them in animal cells which have themselves been cultured, rather like bacteria, in a nutrient medium—a system known as *tissue culture*.

Even with sophisticated modern tissue culture systems, 'cold viruses' are often extremely difficult to isolate and grow from people with the clinical syndrome. Perhaps all viruses that cause the common cold have been identified but methods of isolating them need improvement. Alternatively, there may be other infectious agents, as yet unidentified. Or non-infectious causes of the clinical syndrome called the common cold may be confusing the issue.

In particular, allergies, reactions to chemical pollution, and even certain nervous disorders may result in clinical pictures that are difficult to differentiate from a cold. Often only complex and expensive tests —which involve growing the virus and measuring the patient's immune response to it—can positively identify the causal agents.

Despite these difficulties, a vast amount is now known about cold viruses and the way in which they infect cells. Viruses are minute packages of genetic information which are incapable of reproducing themselves without 'hijacking' the biochemical mechanisms of cells. The smallest are 27 nanometres (0.000000027 m) in diameter, so ten million laid end to end would stretch a distance of less than three centimetres.

By comparison, the cells which viruses infect might be thirty microns (0.00003 m) in one dimension—a million times as large in volume. Similarly, there is between 100,000 and 10 million times as much genetic information in human cells as there is in cold viruses. Yet, in spite of the disparity in size and complexity, just one virus is able to take over the essential functions of such a cell and cause it to manufacture new viruses in overwhelming numbers.

The key point at which viruses take over cell function is protein synthesis. This process normally provides the cell with two vital components: enzymes which control all the chemical reactions taking place in the cell and structural proteins which form some of the building blocks about which the cell maintains its shape. During the viral infectious

The protein coat of the common cold virus attaches to cells so the virus can enter them and protects the genetic material in the centre, which contains all the information needed to make new viruses.

In the nasal cavity, the uninfected cell is protected by both mucus and antibodies in the mucus (A). In (B) a cell infected by a virus is destroyed by an immune killer cell. The cold virus attaches itself to the cell surface, enters the cell and removes the protein coat to reveal the virus genetic information. By inserting a new message (mRNA), virus proteins are produced. Interferon stops the infection by blocking virus protein synthesis (C).

cycle, the machinery of protein synthesis is almost exclusively devoted to making virus proteins rather than cell proteins.

Although the strategies adopted by different viruses for taking over protein synthesis vary in detail, overall their infectious cycles are very similar and are best described in a series of distinct stages.

First, the virus attaches itself to the cell surface. Then it enters the cell and removes the protein coat to reveal the virus genetic information—ribonucleic acid (RNA) or deoxyribonucleic acid (DNA). Next, reading the virus genetic information produces a message—messenger RNA (mRNA)—which takes the place of the cell's own instructions to the protein factory *(ribosomes)*. Hence, virus proteins begin production.

Virus takeover

Virus proteins help complete the takeover of the protein factory and direct the production of more virus messenger RNA and the replication of virus genetic material for inclusion into new infectious particles. The latter process greatly amplifies virus reproduction so that one infecting virus may give rise to thousands of progeny in a single cell.

Virus components, genetic information, structural components and, in some cases, enzymes then assemble into mature virus particles. Finally, the viruses are released from the cell, usually involving the complete destruction, or *lysis,* of the host cell. The free virus particles are now able to infect nearby cells and repeat the cycle so eventually the nasal secretions contain millions of highly infectious particles per cubic millimetre.

Virus infections are not just simple confrontations between the infecting agent and the sufferer's defences. Viruses establish an intimate relationship with normal cellular processes, which makes them very difficult to treat without damaging uninfected cells.

There are essentially three lines of defence against the common cold. The first is physical. Before infection can establish itself, the virus must come in direct contact with a susceptible cell. Not only is the target area small but the nose and throat have a thin lining of mucus which presents a barrier to virus contact. Constantly moving ciliated cells also wash away potential infections.

Mucus itself contains many protective substances. Among them, antibodies form a vital part of the second defence line—the immune system. The immune system consists of a body-wide group of inter-communicating cells which are all essentially related to white blood cells. This network is responsible for first recognizing foreign organisms and then mediating their elimination.

There are, broadly speaking, two pathways by which immune elimination is effected. The first is by the production of special proteins, known as antibodies, which recognize and attach themselves to specific shapes on other molecules. Each antibody recognizes only one unique shape, rather like a lock and key. When attachment takes place the 'recognized' molecule often loses whatever function it previously had. Com-

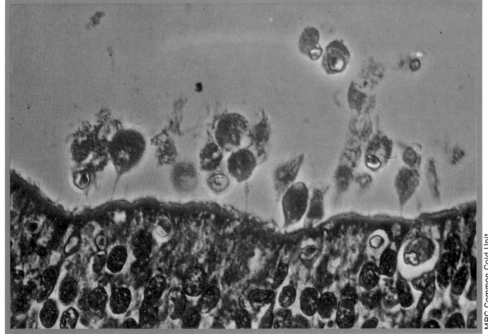

The increase in colds in winter is probably due to a higher rate of transmission. People tend to stay indoors and congregate in warm places such as trains *(above)* where they are likely to infect one another. *Left* A stained section of virus-infected epithelium.

mon cold antibodies are probably most important in inactivating molecules on the surface of the virus, responsible for its attachment and entry into cells.

The second method of immune elimination is mediated by cells which recognize molecular shapes in much the same way as antibodies. These cells subsequently attach themselves to the shape concerned and destroy whatever it is attached to. Since viruses spend most of their time inside host cells it might appear that molecular shapes recognizable to the immune cells are rarely accessible. However, during the course of most infectious cycles, viral proteins are incorporated into the host cell's surface mem-

Medical Science: Virology

brane. The killer cells recognize the viral molecules and kill the infected cells before they can produce a virus.

The immune system is probably most effective in preventing second infections occurring with the same virus. It also plays an important role in limiting the spread of virus and speeding overall recovery.

The final defence system depends on a substance called interferon—a relatively small protein molecule produced in response to any viral infection. Interferon appears to stop the progress of an infection by inhibiting protein synthesis in cells containing a virus.

With such powerful bodily defences, why do viral infections occur at all? At least part of the answer is that the body cannot afford to produce interferon continuously because it inhibits some vital cellular processes such as cell division. This state can be tolerated for short periods, but indefinitely it could to great harm. Blood cells, for example, require cell division in order to function.

The reactions between infected cells, antibodies and 'killer cells' of the immune system lead to the release of various active compounds that circulate in the blood and give rise to headaches, fever and the general feeling of being 'off-colour'. Any factor which influences the state of the defence systems will consequently modify an individual's cold. Such factors include the patient's overall state of nutrition, current hormonal balance and age.

Colds are most commonly transmitted by fine droplets of nasal and throat secretions suspended in the air by coughing and sneezing. The droplets contain large numbers of virus particles, some of which may remain suspended for several hours. Infectious secretions may also be transferred via the hands, crockery or cutlery, for example. However, exposure does not always lead to infection.

Several different cold viruses circulate at any one time. After a certain proportion of individuals within a community have been infected with a particular virus the chances of that virus finding a susceptible individual become remote and the virus apparently disappears from that community. Its place is

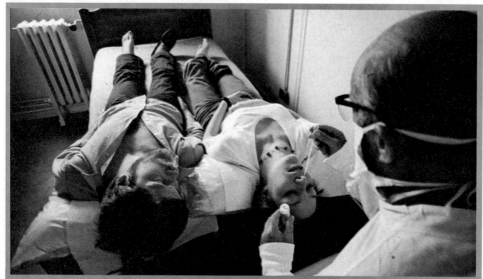

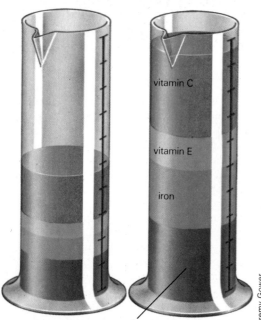

COMPARISON OF IMMUNE FUNCTION

zinc, copper and many other trace elements

For optimum immune functions *(left)* all these substances are needed, perhaps in the proportions shown. Vitamin C is important in cell movement; the others are essential components of enzymes controlling cell function. Deficiency of any one or all of them *(far left)* reduces immunity. At the Common Cold Research Unit, new volunteers are briefed *(top)* before the cold viruses are administered *(above). Right* Cell cultures are being prepared.

usually taken by another virus type to which members of the community are not immune. Thus, a network of communities within which rates of infection with given viruses are waxing and waning according to the group immunity could be built up.

All non-prescription drugs treat the symptoms of the common cold—thereby making the illness less disagreeable—but they have no effect on the growth of cold viruses. Most are combinations of antihistamines or adrenaline-like drugs (which dry up nasal secretions) and aspirin or paracetamol (which eases aches and lowers the temperature).

Destroying viruses

Treating the cause of the common cold—and indeed any viral infection—is one of the great challenges of modern medicine. The intimate relationship between virus and infected cell makes it hard to destroy viruses without destroying normal cells as well. The problem is like that of treating cancer.

Many agents have been found to inhibit virus replication but frequently they turn out to be toxic to normal cellular processes. Where the infection is potentially fatal it may be worth the risk of using such agents, but in a trivial infection such as a cold this approach would be too drastic.

Recently, an apparently safe drug has been found to inhibit the multiplication of rhinoviruses. In trials in 1981, it appeared to be moderately effective in alleviating the symptoms. The drug, Enviroxime, is ad-

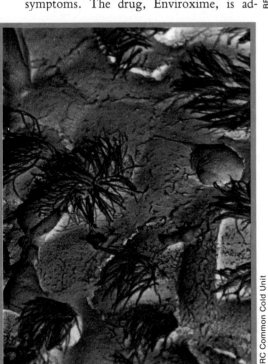

Above Spraying buses: an anti-flu measure of the 1920s. *Left* Nasal epithelium attacked by a rhinovirus. *Right* Eskimos are particularly vulnerable to cold viruses because they lack immunity—not because of their cold climate.

ministered both orally and into the nose.

Enviroxime's mechanism of action is not yet clearly understood, but it appears not to affect virus entry into cells. It may be of value to patients who are likely to suffer the more serious consequences of colds.

Even more recently, a compound called dichloroflavan has been found to inhibit rhinovirus RNA synthesis in cell cultures. The simplicity of this substance and its close relationship to known non-toxic substances may make it useful in the future.

At present, interferon is far too expensive and valuable for use in combatting the common cold although it is known that it can both protect people if given before exposure and curtail the infection if given when symptoms develop. At one time it was hoped that drugs could be developed which induced local interferon manufacture in the nose. Such drugs exist, but their action appears to be too slow for use during acute infection.

With around 140 different cold viruses, vaccination is not feasible either. Even if all the vaccines could be made, people would be unlikely to have all the injections.

Many vitamins and trace elements are known to optimize the immune system's ability to deal with infections. So, with an ever-greater understanding of optimum human diets—as well as improved living standards—we can reasonably expect our immune systems to become better equipped to cope with the common cold virus. The prospect of discovering a drug that is applicable to all cold sufferers remains more remote.

Technology Tomorrow: Solar Power

Above The Solar Challenger—not an all-weather aircraft perhaps, but its historic Channel crossing in 1981 did highlight the incredible potential of the photovoltaic cells for long-term development.

Cell of the century?

Holidaymakers lying on a sunny beach cannot fail to be aware of the existence of solar energy. But when they return home and switch on their television sets, their holidays over for another year, solar power still affects them. For the foreign programmes they enjoy would stay firmly in the transmitting stations were it not for the solar panels which power the satellites passing the signals around the world. Such is the power of photovoltaics.

The photovoltaic effect is a process in which two dissimilar materials pressed into tight contact act as an electric cell when struck by certain types of light.

Any semi-conducting materials can be used to demonstrate the effect, but silicon is the element most commonly employed because in addition to being a fairly efficient converter of light into electricity it is both cheap and plentiful.

A typical silicon solar cell consists of two layers of silicon, each 'doped' with different impurities. On top is a network of grid contacts which constitute the negative contact, while beneath the cell is another grid network forming the positive contact.

Free electrons

Electrons do not normally move from atom to atom within the silicon crystal, but when light strikes the crystal this provides the energy needed to free some electrons from their bound condition. These free electrons cross the junction between two dissimilar crystals more easily in one direction than the other with the result that one side of the junction is given a negative charge with respect to the other, in the same way as a conventional battery.

By wiring a circuit between the positive and negative contacts of the cell, the electricity produced can be used normally, and will continue to flow for as long as light falls on the two materials.

The amount of energy needed to release the electron is called the *band gap*. This is expressed in the form of electron volts, and corresponds to the wavelength of light which has that amount of energy. Because the band gap varies from one type of semi-conductor to another, each type responds to a different type of light.

The vital part of the cell is the junction (sometimes known as the barrier layer) between the two materials. In the silicon cell, the conventional way of producing the junction is by *doping* the silicon.

This is a process by which small quantities of impurities are added to the crystal lattice so that positive ions are implanted on one side of the cell, and negative ions on the other. Wafers of silicon are baked in an oven in a gaseous atmosphere which provides the required impurities. In these conditions, individual atoms diffuse into the crystal lattice and it is these impurities which permit free electrons to flow under appropriate conditions. In a typical cell, the two layers are doped with phosphorus (which produces the negative n-layer) and boron (which produces the p-layer.)

Individual cells produce very low power outputs at the best of times so to produce usable amounts of electricity the cells are manufactured in panels. Each panel consists of a number of cells connected together to produce a known voltage output. Once this has been done it is possible to provide any power at any voltage—the voltage being doubled by connecting two panels in parallel.

In theory then, there appears to be no limit to the output of solar cells; but in practice the vast majority of solar converters are restricted to the 1 kW range and there is little prospect of this being bettered until the efficiency of cells in converting light into power is greatly improved.

The theoretical maximum efficiency of a silicon solar cell is 45 to 50 per cent, but this is with monochromatic visible light—either a pure yellow or green beam. For practical applications sunlight must be the light source; this is low in efficiency because the solar spectrum extends from ultraviolet to infra-red light, and silicon cells are insensitive to light outside the visible spectrum and a very small part of the infra-red spectrum.

The maximum conversion efficiency of silicon cells operating outside the atmosphere—those powering a satellite, for example—is around 19 per cent. Strangely, although light intensity on land is lower than in space, the maximum conversion efficiency rises to around 23 per cent at sea level because the different spectral composition of light inside the atmosphere has the effect of concentrating more of the incoming energy in the parts of the spectrum where solar cells are sensitive. However, current technology is a long way from approaching these maximum theoretical effeciencies and most current commercial cells have an efficiency of between 10 and 16 per cent.

Sunny South Pole

The power output of solar cells is inevitably limited by the amount of sunlight falling on them. For this reason, the application of photovoltaic cells in temperate areas of the world will always be more limited than in, say, the tropics. But it is important to note that the photovoltaic effect does not depend upon heat, but only on light. In fact, the efficiency of solar cells drops as the temperature rises; solar cells have been used at the South Pole and these delivered more power than would be expected in a typical temperate climate because the light was able to transfer its energy directly to the electrons in the cells without losses that would be caused by internal resistance.

The output of solar cells can be increased by reflectors which concentrate more light on the panels. One simple method of doing this is by siting the cells in a trapezoidal groove with mirrored walls. So long as the concentrator is installed on an east–west axis so that the mirrors do not obstruct the Sun's rays, no further orientation to the Sun is required and anything up to a fourfold improvement in output can be achieved.

Even better is a compound parabolic concentrator—also known as a Winston collector—which can show improvements up to a factor of ten; but this adds a further complication in that it needs periodic north–south tilt adjustment to keep it properly oriented to the Sun.

It is possible to get a concentration up to a factor of around 5,000 by use of parabolic reflectors, but these would require very accurate tracking, water cooling to protect the solar cells, and the use of a different form of cell because silicon units rapidly lose efficiency at high temperatures.

One cell that might be used with a concentrator is the Gallium Arsenide (GaAs) solar cell; not only does this have good conversion efficiency, but it also continues to work well at high temperatures—above 100°C, the GaAs cell performs better than conventional silicon units. It was for this reason that the USSR used GaAs cells on its lunar projects, where daytime operating temperatures are high.

Another possible solar cell is the Cadmium

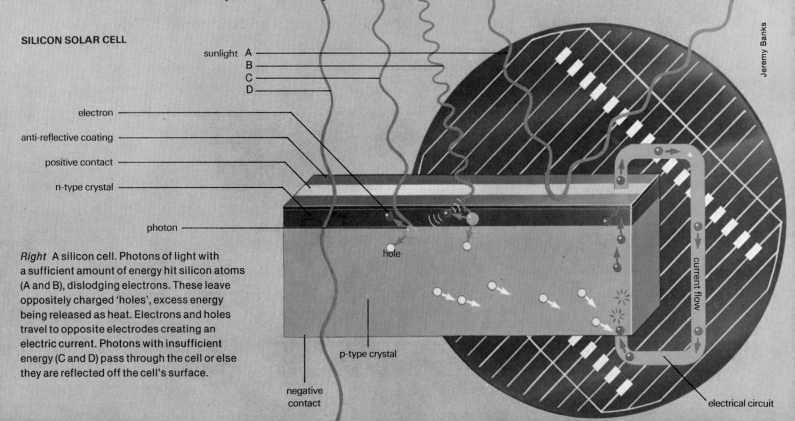

SILICON SOLAR CELL

Right A silicon cell. Photons of light with a sufficient amount of energy hit silicon atoms (A and B), dislodging electrons. These leave oppositely charged 'holes', excess energy being released as heat. Electrons and holes travel to opposite electrodes creating an electric current. Photons with insufficient energy (C and D) pass through the cell or else they are reflected off the cell's surface.

Technology Tomorrow: Solar Power

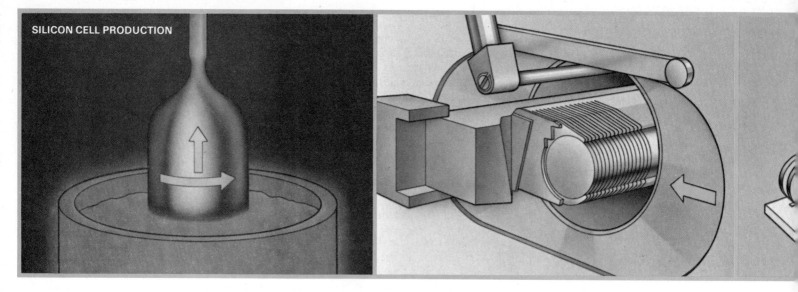

SILICON CELL PRODUCTION

Above In the Czochralski method of growing silicon crystals a rotating seed crystal is dipped into molten p-type silicon and then it is slowly withdrawn *(below)*. It is sawn into wafers and doped with phosphorus *(above centre* and *above right)* before the electrical contacts are attached to it. A non-reflective coat is added to completed cells *(below right)*.

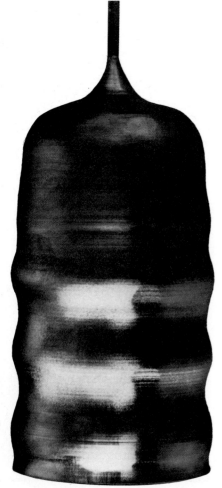

Sulphide (CdS) unit, which is low in efficiency but has the great advantage of cheapness. For this reason it is being looked at as a serious possibility for large-scale plants. Although CdS cells were used from 1967 to 1973 for powering French stratospheric balloons, their low efficiency and poor power-to-weight ratio prevent their future use in similar applications.

Other types of cell still under development include the Indium Phosphorus (InP) and Cadmium Tellurium (CdTe) cells, with maximum efficiencies of 23 and 25 per cent.

The greatest advantage of all photovoltaic cells is that they require virtually no maintenance or supervision; once they are in position and the system is set up, they can practically be ignored.

Nowhere are these qualities more appreciated than in space, and solar cells have proved themselves to be eminently suitable for such applications, being used on practically every space vehicle. Enormous weight savings can be made because it is unnecessary to carry any fuel for the power source. Even when large quantities of power are needed, space agencies still look to the solar cell because it is lighter than any other power source; Skylab, launched in 1973, required a 20 kW system, and this has been provided by photovoltaic cells.

On Earth, weight tends not to be the overriding factor and for photovoltaic cells to be a practical proposition, they still have to show other advantages over alternative power sources. The fact that practically no maintenance is required was obviously a factor in favour of photovoltaic cells when, for

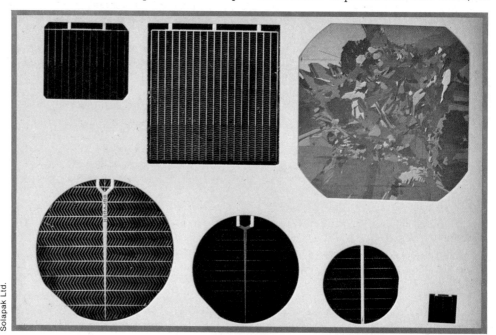

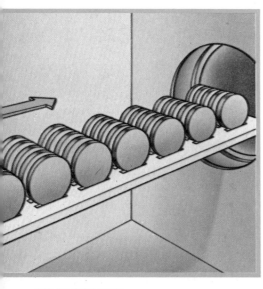

SOLAR CELL TYPES

Cadmium Sulphide (CdS) cell

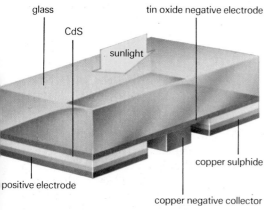

Gallium Arsenide (GaAs) cell

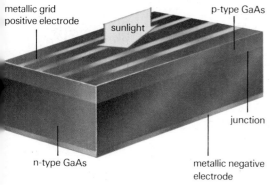

Top The Cadmium Sulphide (CdS) cell is very cheap but not very efficient. The Gallium Arsenide (GaAs) cell *(above)* remains efficient at high temperatures, making it an excellent unit for use in space activities. But the silicon cell is the most widely used for powering (among many other things) telephones *(right)*, due to its good all-round performance.

example, several unmanned navigational lighthouses were built in England and a 360 kW landing beacon was erected in the remote area of Anchorage, Alaska.

An interesting diversion can be seen in Miami, Florida, where the local telephone company has one of its public phone booths powered by a solar panel on the roof. However, this is little more than a publicity stunt and cannot be seen as a serious photovoltaic application. Similarly, the Florida Power and Light company spent over half a million dollars in early 1981 installing a photovoltaic system to run the air-conditioning system for their head office. But air-conditioning requires very high power outputs and is therefore not a suitable application for solar power—which is why the system cost so much to put in and why it cost far more than a conventionally powered system would have done.

For most applications, the spread of solar cells is limited by the fact that they have to compete in cost with existing power sources, and by the fact that the panels themselves will take up a not inconsiderable amount of space in anything but modest low-power installations.

Solar irrigation

In tropical and equatorial zones, however, the prospects are far brighter because of the longer hours of sunlight and the greater intensity of that light. Already, solar plants are in quite widespread use in these regions, mainly for applications requiring a steady source of power where other methods would be expensive or impracticable.

It is not unusual to find solar panels in use for pumping and irrigation purposes in rural areas; all that is required are the panels, which can be placed anywhere in the Sun, a number of lead-acid storage batteries and, if an AC (alternating current) supply is required for the pump, a simple inverter.

The sticking point of many possible installations is cost. Where a photovoltaic system may be perfectly feasible, few people are likely to install one where other forms of power are cheaper. But if the capital is available then solar power becomes very attractive—as the Saudi Arabians found when they covered literally acres of desert with solar panels and used the electricity supply to power a village nearby. Where a power station already exists, however, it may still be cheaper to construct power lines.

Solar cells remain expensive not only because they are produced in relatively small quantities using very high technology. Quite apart from these factors, it is necessary in the interests of cell efficiency to ensure that the materials used are extremely pure.

A lot of research has gone into reducing the production costs of solar cells. In many instances this takes the form of experimentation to find semi-conducting materials which can be used in cells of reasonable efficiency, but which do not require high, and therefore expensive, purity. Another way of bringing down production costs would be to select materials and technologies that are suitable for a continuous production process.

In both these respects, the CdS cell looks promising for the future because the materials used do not need to be of extremely high purity and because it may well be found possible to produce such cells continuously using an adaptation of the 'float glass' production technique.

Another encouraging possibility has been mooted by Dutch researchers concerned with reducing the cost of 'doping' the semi-conducting materials. Instead of using the costly and time-consuming oven baking methods, they are looking to a laser beam to

Above right The efficiency of solar cells depends on their band gap. A solar-powered navigation beacon in England *(left)* easily uses its solar array to charge 10 storage batteries and power the light. Battery storage capacity is about 65 amp/hours.

implant the required ions into the crystal surface. And because of the ease of automation that this would bring, it is estimated that this process alone might reduce the cost of solar cells to two per cent of their present cost making them very competitive.

Similarly, Boeing Aerospace have developed a thin-film solar cell based on Copper-Indium Selenide and Cadmium Sulphide; although its conversion efficiency is only around 9.4 per cent, its production cost per watt of power produced is claimed to be around five per cent of the production cost of conventional silicon cells.

As soon as production costs come down, there is no doubt that solar cells will be far more widely used—after all, their energy source is both free and plentiful. Even now, it is more problems of availability than cost that prevent integrated solar power systems being widely adopted for domestic use in the developing world.

In the West Indies, for example, a homeowner could install three simple panels on the roof of his house and connect these up to two deep-cycle 225 amp/hour lead-acid batteries; with the panels rated at 1.5 amp/hours each, and producing 12 volts DC, this would provide adequate power for four or five lights, fans, a small water pump, TV, radio and other domestic appliances. Because of its high energy requirements, a freezer or air conditioning unit would need further panels, but the whole system could be installed for very little more than the cost of the only alternative power source—a diesel generator. And once the solar system is installed, no further fuel would have to be paid for and imported from outside.

Transport is another area in which the

Technology Tomorrow: Solar Power

solar cell could play a part. Already a motorcycle has been produced with a range of 65 km (40 miles) and a top speed of 50 km/h (30 mph). One very interesting concept is for a small car with photovoltaic panels covering the roof. These cells would produce a significant charge all through the daylight hours, putting around 12 amp/hours into the lead-acid batteries that would fill the 'engine' compartment and boot. Quite apart from providing personal transport, it is conceived that the owner's house should be wired up to the car's storage batteries when the owner is at home so that all the internal domestic systems would run off it as well.

Already, photovoltaic cells have been used to power low-energy devices such as watches and calculators—though usually in conjunction with a nickel-cadmium battery—and there is little doubt that many more examples will be introduced in the near future. The medium got a great deal of publicity when a totally solar-powered aeroplane was flown across the English Channel, so it is not wholly inconceivable that with improvements in cell efficiency, solar-powered commercial aircraft may one day be seen.

But these are really little more than sidelines; where photovoltaic cells score over all conventional power sources is that they are clean, non-polluting and noiseless. That in itself should be good enough reason to see their far wider application in all areas of modern life. Add to that their low maintenance, economy, reliability and simplicity and it seems inconceivable that photovoltaic cells will not be given the research and applications that they deserve.

Above Concerned with the world's dependence on fossil fuels, Saudi Arabia has now built a solar-powered village near Riyadh. Martin-Marietta in the USA both built and installed the 160 solar arrays *(inset)* which provide the area with 350 kW of power. The map *(below)* shows the potential for efficient daily usage of solar cells in different parts of the world. The solar cells sited on the roof of this car *(left)* have not replaced its fuel tank—but the electricity that they produce does reduce its fuel consumption by five per cent.

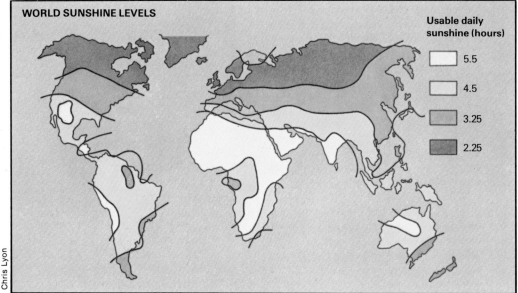

WORLD SUNSHINE LEVELS

Usable daily sunshine (hours)
- 5.5
- 4.5
- 3.25
- 2.25

2329

Resources: Materials

Leather goes from strength to strength

Leather, the first material Man used for clothing, is still high fashion today despite being a luxury option rather than the sole alternative to leaves. Although experiments in leather processing have included using atomic radiation for preserving skins, the leather industry tends to be conservative in its methods—with good reason. Since no two animals eat the same food, no two skins are exactly alike, and so no machine can equal human judgement and dexterity at certain vital stages of leather manufacture.

The fact that each piece of leather is unique presents problems in manufacture, but it is also one of the main reasons why people find satisfaction in possessing leather apparel and goods. And no synthetic substance yet invented can offer the same combination of practical qualities as leather. Thus it remains the staple raw material of the footwear trade and is equally indispensable in many advanced technical applications.

Although leather is presented to the public as a highly desirable commodity, it is actually a by-product, or even a waste product, of the meat industry. Very few animals are kept for their skins alone so the quality and quantity of raw material available depends on the current state of the meat market.

Hide and skin

The modern demand for leaner meat has led farmers to breed animals with less fat—making their skins much harder to remove—and the economic recession is leading to decreased herds. Nevertheless the leather industry is holding its own by tapping fresh resources for skins and by employing technology such as the Japanese machine which triples the yield from a single skin by slicing it into three fine layers.

All leather is the skin of animals when it is

No tannery smells sweet but only a few have a stench to rival the one found in the walled city of Fez, Morocco *(right)*. The smell is due to the urine which is incorporated into the traditional curing mixture that is still used. Despite the smell, the tanners' *souk*—or bazaar—is a tourist attraction. There exists a kind of leather known as *morocco* because it was first made by the Moors. It is a soft goatskin, often used in bookbinding.

tanned or preserved to prevent natural bacteriological deterioration but the industry only uses the term *skins* for smaller animals—goats, sheep, calves or reptiles. The skins of cattle, buffalo and larger animals are always referred to as *hides*.

To understand the processes of leather manufacture it is essential to look at the make-up of skin. This has four layers: the outer layer is the *grain*, which includes the epidermis and sebaceous glands; the second contains the veins, arteries, hair follicles and roots; the third is the *corium*, a network of interwoven collagen fibres; the fourth is the flesh side of the corium containing fatty tissue and very coarse collagen fibres.

In the living skin the collagen fibres are contained in a watery jelly of a protein-like substance. When the skin is dried, this substance becomes a hard, glue-like material, cementing all the corium fibres so that the skin becomes hard and horny. Consequently the inter-fibrillary protein has to be removed if the leather is to be soft and supple.

As no individual animal has the same diet or grows at the same rate, the fat and protein levels in each hide or skin differ. Manufacturers require even colour, texture and body, so most of these natural substances have to be removed and replaced with *tanning agents* (preservatives) and *fat-liquors* (oils or fats) as well as *dyes* (colouring).

To make smooth-grain or *nappa* leather the epidermis is removed from the first layer to leave a pattern of hair follicles which is different in each species. The leather may then be used with its natural grain effect or finished with cellulose coatings or pigments. Patent leather involves adding lacquer at this stage. Suede leather results from buffing the ends of the collagen fibres from the fourth layer—or, if the skin is split, from the underlayer of the split.

Preserving skins between the abattoir and tannery has been the subject of much recent research. Atomic radiation has been tried out in the USSR as a means of neutralizing raw or *green* cattle hides. Starting with small doses of radiation on a few hides, it was eventually found possible to preserve a pallet load of hides weighing one tonne in ten minutes, with a lasting effect of six days. Unfortunately this line of research had to be abandoned because the hides retained such a high level of radiation that they were dangerous for tannery workers to handle. Also, making the neutralizing plant safe necessitated using insulating materials equivalent to a depth of five metres of water. The scientists involved in this project are now investigating the use of accelerated neutrons as a safe alternative process to pure radiation.

The US meat industry finds the quickest and most effective method of curing to be *brining*—green hides are run in large drums

Above left The market at Nabul in Tunisia caters mainly for the tourists. The local leather specialities include goods made from camel hides and goatskin. *Far left* The International Leathermark was adopted in 1973 to supersede the many national marks. It identifies articles made wholly or partly from leather. *Left* Animal skin has distinct layers of different substances.

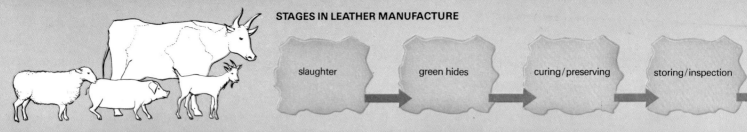

STAGES IN LEATHER MANUFACTURE

slaughter → green hides → curing/preserving → storing/inspection

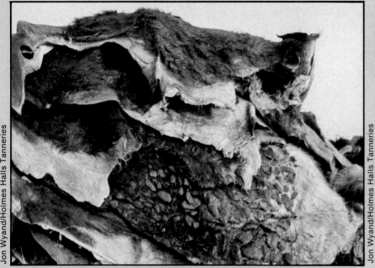

in a salt solution combined with a few preservative chemicals. If carried out within one hour of the hide being removed from the carcase, this process will stop any deterioration in the quality of the hide or skin. It is also quite possible to preserve hides and skins adequately by cooling them to around 0°C. Freeze drying has been tried but requires an expensive plant, as does keeping hides at low temperatures for prolonged periods.

At the tannery

The leather industry is attempting to educate farmers and butchers, particularly in the Third World, to look after their animals and also the hide once it has been taken off the carcase. The United Nations is actively involved with the governments of most Third World countries in teaching simple curing methods and setting up systems that aid the recovery of hides and skins which are presently either burnt, buried, or eaten.

To make leather from raw skins and hides the tannery takes a number of steps which differ in detail, depending on the particular finished product required. First the hair or wool is removed using depilatory chemicals. This process—called *unhairing*—is one of the two major sources of pollution in leather-making. The strength of the chemical depends on the type of hide or skin and the fat or grease content as well as on the value of the hair or skin being removed. The most common method of unhairing is to paint the flesh in a wash of hydrated lime, sodium

Above left The starting point: salt-cured raw bovine hides. They are immersed in lime drums for one day for unhairing, and then fed into the fleshing machine *(above right).*
Right Two skilled operators are needed to feed a wet hide into a splitting machine, and three strong ones are needed to pull it through.

sulphide and water. The sodium sulphide enters the corium from the flesh side and dissolves the keratin cells enclosing the hair roots. The skin is then unhaired from the grain side either by hand, over a 'beam' using a curved knife, or by a machine which has a revolving knife cylinder. Care has to be taken not to damage the surface layer of the skin with the knives or to overexpose it to the chemicals.

Unhairing may also be carried out by immersing hides or skins in a mixture of lime and water or by running them in a drum containing water, sodium sulphide and salt. But over-liming can burn the surface layers of the grain side and also loosen the fibres of the corium to such an extent that the hide or skin will tear easily. Other chemicals used in unhairing include sodium hydrosulphide, calcium hydrosulphide and arsenic sulphide and these produce extremely toxic waste waters. The leather industry is currently researching ways to reduce this type of environmental pollution.

After the astringent process of unhairing, the hide or skin is turned over and *fleshed*—the removal of any flesh still adhering to

SPLITTING MACHINE CROSS-SECTION

endless band knife, support roller, grain, gauge roller, hide, section roller, support roller, flesh split, outlet plate

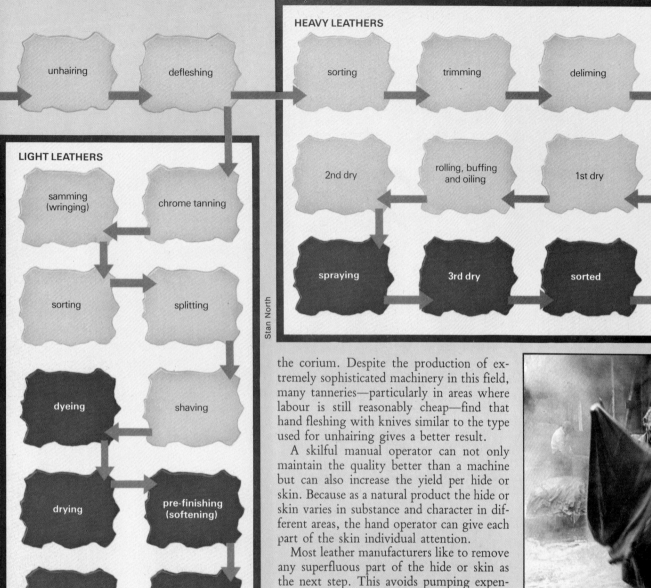

the corium. Despite the production of extremely sophisticated machinery in this field, many tanneries—particularly in areas where labour is still reasonably cheap—find that hand fleshing with knives similar to the type used for unhairing gives a better result.

A skilful manual operator can not only maintain the quality better than a machine but can also increase the yield per hide or skin. Because as a natural product the hide or skin varies in substance and character in different areas, the hand operator can give each part of the skin individual attention.

Most leather manufacturers like to remove any superfluous part of the hide or skin as the next step. This avoids pumping expensive chemicals and dyes into material that will later be discarded. Similarly, hides are often split at this stage. Sophisticated automatic substance levelling and splitting

Left Hides suspended in tanks containing wattle tannage, which takes eight days to fully penetrate the hide and also chemically react with it to make leather. *Above* Hides being removed from dye drums which act like giant washing machines; each of these drums holds 100 hides. This is full-grain shoe upper leather, which needs five hours of saturation with chemical dyes. Processes vary according to the type and weight of the raw hide and the end product that is desired. The processes pictured here are for shoe leather—soles and uppers—produced at the UK's largest tannery, Holmes Halls, which processes at least 2,800 hides per day.

Resources: Materials

machines have been made in recent years. With a minimum of skill hides can be now split into two or more parts, while being kept even in substance and close to the thickness required for the finished article. The most valuable split is the grain side: this can be finished into high-quality grain or suede leather, depending on which side of the split is used. The flesh split is generally of lower quality and is used for cheap industrial clothing or low-grade shoe uppers. The main problem with the flesh split is that the splitting knife cuts through the centre of the corium. The fibres on the flesh side are coarser and do not have the same resistance to tearing as the grain split. It is not usual for skins to be split at this stage as they are much more tender than hides and it is far easier to split them in a tanned or semi-finished state.

Pollution problems

The next stage is to *tan* the hides and skins, at which point they actually become leather. Thirty years ago most tannages were based on the barks of trees such as oak, wattle, quebracho or 'vegetable' tannages. These were extremely good and had their own particular characteristics but, in the search for increased softness, experiments proved that tanning leather in chromium salts resulted in extremely soft and light-fast leathers. Chrome-tanned leathers were also found to be more heat resistant, easier to waterproof and in some cases could even be made fully washable.

Needless to say, the new discovery had its drawbacks. It was found that heavy metals were undesirable pollutants of the countryside and were not bio-degradable, though in the past five years there has been much research into treating waste chrome liquors and finding adequate substitutes. Recent studies have investigated the use of sodium aluminium silicate, aluminium salts combined with polycarboxilic high polymers, and (in the USSR) titanium and zirconium.

At the time of tanning, fat-liquors have to be added in order to replace the natural oils and grease removed during the earlier processes. These fat-liquors caused major problems for the leather industry for many years when sperm whale oil was used as a softener for gloving and garment leathers. Today whale oil substitutes are almost universally used. At the fat-liquoring stage, considerations such as light-fastness have to be taken into account again. It has recently been found that the addition of aliphatic sulphonamides during fat-liquoring considerably improves the light-fastness quality

Above Shoe sole leather. *Right* The appliqués of several different textures of leather and stencilling form these Art Deco designs.

that can be obtained in finished leathers.

At this stage the lighter leathers from skins are levelled in thickness. They can be shaved on the flesh side, with machines containing cylindrical knives, in either a damp or dry state; it is more common to wet-shave as there is less danger of overheating the skins or causing excessive wear to the machine. Most developments in this area involve simplifying the operation of the shaving machines—an unskilled operator can tear skins to pieces very quickly on a traditional model of shaving machine.

Bovine hides, by contrast, follow the production line straight from tanning into the dyeing stage without being shaved unless

further splitting is required. Some of the latest techniques in splitting have been developed by the Japanese. They can split even delicate skins like wool sheepskins down to a thickness of 0.2–0.3 mm. They then back the split with very fine silk to restore the tensile strength lost during splitting. This allows three pieces of leather to be made out of the same skin.

At the dyeing stage the final colour is determined. Although it is always possible to change the shade of grain leather by adding pigments or top coats to the grain side, a suede's colour depends entirely on the dyeing. The operation is traditionally carried out in large wooden drums, or more recently in stainless steel processors which rotate the hides very much like washing machines. Stainless steel processors have a high capacity, rapid operation cycle and use far less water than the traditional drum, thus reducing the volume of waste waters that have to be treated and cleaned before discharge into public waterways and sewers.

Fashion now decrees that leathers look natural and be as soft and supple as possible. This has led to the widespread use of aniline dyes which give the leather colour but leave its natural characteristics visible. However, these dyes are not water resistant and the leather has to be further treated if the garment or shoe is not to be affected by rain spotting. It is possible to put semi-transparent topcoats over aniline dyes but only at the expense of losing some of the natural beauty of the hide or skin.

Right It takes 60 hours to make a custom-built saddle at Bridleways of Guildford in the UK, although their master saddler *(above right)* has over 55 years experience. Pigskin is used for the seat and the strong parts of the cowhide, termed the *butt,* for the flaps and skirt. *Below right* The saddlery tools have hardly changed in design in over 150 years.

LEATHER GOODS: THE CUTTER'S CHOICE

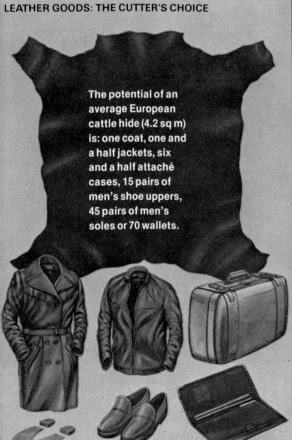

The potential of an average European cattle hide (4.2 sq m) is: one coat, one and a half jackets, six and a half attaché cases, 15 pairs of men's shoe uppers, 45 pairs of men's soles or 70 wallets.

Resources: Materials

Some leathers such as patent leather are screen-coated or sprayed with heavy laquers or pigments which are often cellulose-based. Most grain leathers are polished or glazed in their final process. Suedes which have been dyed are buffed or wheeled on machines containing drums covered in emery paper to bring up the nap; they then have to be carefully brushed to remove the resulting dust.

It is a common fallacy that man-made fibres and substitutes have made considerable inroads into the traditional areas in which leather has been used. About 20 years ago, when the shoe upper material Corfam was introduced to the market, this seemed highly likely. However, its price caused its withdrawal after only a few years. Most substitutes for leather are made from materials based on the petro-chemical industry. Their advantage is that they can be produced in long runs of uniform size, width or substance which makes it possible for the finished product manufacturer to maximize the cutting from a given length of material. Hides and skins vary in size and shape and if cutters are not skilful or carefully supervised, profit margins can often literally be thrown into the waste bin.

The overriding advantage of leather, however, is that so long as Man eats meat—and often even if he does not—there will always be hides and skins available. While the petro-chemical industry has very finite resources, the vast resources of raw material for the leather industry have hardly been touched.

Culling raw material

The Third World has the most potential for both culling raw material and processing it. India has already taken advantage of this situation, using trade embargoes to push its leather industry rapidly ahead. At the opposite end of the scale, many Western European tanneries have closed down in recent years because of the reduced spending power of the average European. The tanneries that have survived have had to concentrate more than ever before on very high quality combined with high technology finishes that produce a variety of speciality items that are suitable for a sophisticated market.

The uses to which leather is put are very varied. Apart from handbags, briefcases, wallets and belts, it is now commonly accepted for outer garments and, using new processes, leather clothes such as shirts that can be washed in a washing machine in the same way as textiles. Even the world of advanced technology still uses leather. Parts of the flight deck of the British Aerospace's 1–11 aircraft have a number of leather components. Gasmeters contain a diaphragm made of very thin, flexible, gas-proof leather and no substitute for gasmeter leather has yet been perfected. The flexibility and durability of leather makes it ideal for the bellows, puffer motors and the keyboard bedding of church organs.

From cave to space

Other traditional uses include leather washers and oil seals, and a recent boost has come from the resurgence in the use of the heavy working horse: no synthetic material can equal it for either looks or wear in harness or saddles. Woolled sheepskins have recently come into demand—hospitals use them for preventing bed sores in bed-ridden patients and airlines find they prevent fatigue when used for pilots' seats. All in all, leather has never failed to keep pace with humans at every stage of their progress from the cave to the space station.

Leather still reigns supreme in many sports. *Above* Its main advantages for motorcycle gear are good protection against injury, warmth from one layer, and streamlining. *Left* Its durability scores high in cricket.